A CULTURAL HISTORY OF CHEMISTRY

VOLUME 2

A Cultural History of Chemistry
General Editors: Peter J.T. Morris and Alan J. Rocke

Volume 1
A Cultural History of Chemistry in Antiquity
Edited by Marco Beretta

Volume 2
A Cultural History of Chemistry in the Middle Ages
Edited by Charles Burnett and Sébastien Moureau

Volume 3
A Cultural History of Chemistry in the Early Modern Age
Edited by Bruce T. Moran

Volume 4
A Cultural History of Chemistry in the Eighteenth Century
Edited by Matthew Daniel Eddy and Ursula Klein

Volume 5
A Cultural History of Chemistry in the Nineteenth Century
Edited by Peter J. Ramberg

Volume 6
A Cultural History of Chemistry in the Modern Age
Edited by Peter J.T. Morris

A CULTURAL HISTORY OF CHEMISTRY

IN THE MIDDLE AGES

VOLUME 2

Edited by Charles Burnett and Sébastien Moureau

BLOOMSBURY ACADEMIC
LONDON • NEW YORK • OXFORD • NEW DELHI • SYDNEY

BLOOMSBURY ACADEMIC
Bloomsbury Publishing Plc
50 Bedford Square, London, WC1B 3DP, UK
1385 Broadway, New York, NY 10018, USA
29 Earlsfort Terrace, Dublin 2, Ireland

BLOOMSBURY, BLOOMSBURY ACADEMIC and the Diana logo are trademarks of Bloomsbury Publishing Plc

First published in Great Britain 2021
Paperback edition published in 2025

Cover design: Rebecca Heselton
Cover image © Bibliotheque de L'Arsenal, Paris, France/Archives Charmet/ Bridgeman Images

A catalogue record for this book is available from the British Library.

A catalog record for this book is available from the Library of Congress.

ISBN: PB: 978-1-3505-5204-3
Pack: 978-1-3505-5229-6
ePUB: 978-1-3502-5149-6
ePDF: 978-1-3502-5148-9

Series: The Cultural Histories Series

Typeset by Integra Software Services Pvt. Ltd.
Printed and bound in Great Britain

CONTENTS

List of Illustrations vii

List of Tables xi

Series Preface xii

Introduction 1
Sébastien Moureau

1 Theory and Concepts: The Shared Heritage of Byzantine, Arabo-Muslim, and Latin Alchemy 17
Matteo Martelli, Sébastien Moureau, and Jennifer M. Rampling

2 Practice and Experiment: Alchemical Operations in the Middle Ages 35
Sébastien Moureau and Nicolas Thomas

3 Laboratories and Technology: Alchemical Equipment in the Middle Ages 49
Nicolas Thomas and Sébastien Moureau

4 Culture and Science: Alchemy's Scientific Contexts and Critiques 71
Regula Forster and Jean-Marc Mandosio with Antoine Calvet and Gabriele Ferrario

5 Society and Environment: The Social Position of the Alchemist and Alchemy in the Court, in the Church, and in Society 93
Charles Burnett with Antoine Calvet and Justine Bayley

6 Trade and Industry: Medieval Craftsmanship and Technology Transfer 107
Justine Bayley with Spike Bucklow

7 Learning and Institutions: Teaching the Art East and West 131
Regula Forster and Jean-Marc Mandosio with Antoine Calvet

8 Art and Representation: The Alchemical Image in the Islamic and Christian Middle Ages 149
Jennifer M. Rampling

Notes 179

Bibliography 185

List of Contributors 213

Index 214

LIST OF ILLUSTRATIONS

2.1 Schematic representations of alchemical processes. 1. Distillation *per ascensum*. 2. Distillation *per descensum*. 3. Rectification (repeated distillation). 4. Cohobation or circulation (circular distillation). Drawing by Nicolas Thomas 41

2.2 Schematic representations of alchemical processes. 1 & 2. Sublimation. 3. Descension. 4. Refining by cementation. Drawing by Nicolas Thomas 45

2.3 1. Schematic representation of cupellation. 2. Schematic representation of fire assay. Drawing by Nicolas Thomas 46

3.1 Alchemical apparatus. Bibliothèque nationale de France, Paris, MS Lat. 7162 (fifteenth century), fol. 92v 53

3.2 Aludels, alembic, furnace, and other alchemical equipment. University of Pennsylvania Library, Philadelphia, Edgar Fahs Smith Mem. Coll., MS Codex 69 (ca. 1450–1475), fol. 22v 54

3.3 On the left, fractional distillation. On the right, cohobation or circulation. *Ordinal of Alchemy* of Thomas Norton, British Library, London, Add. MS 10302 (1477), fol. 37v 54

3.4 Archaeological finds of alembics for distillation *per ascensum* (pottery: 1–3 & 6, glass: 4 & 5). (1) Paris (France), Louvre, mid-fourteenth century (drawing by Stephen Moorhouse in Rouaze 1986). (2) Basel (Switzerland), Ringelhof, mid-thirteenth century (Kamber and Kurzmann 2002). (3) Marseille (France), Bourg des Olliers, Sainte-Barbe, thirteenth century (Vallauri and Leenhardt 1997). (4 & 5) Wakefield (UK), Sandal Castle, first half of the fifteenth century (Moorhouse 1983). (6) Köszeg (Hungary), late fifteenth to early sixteenth centuries (Holl 1982) 57

3.5 Archaeological find of two pots luted in a pit for distillation *per descensum* (pottery). Paris (France), Louvre, fifteenth century (drawing and photography by Catherine Monnet in Thomas and Claude 2011) 58

3.6 Archaeological finds of aludels for sublimation (pottery). (1–4) Beaucaire (France), Abbey of Saint-Roman-l'Aiguille, fourteenth century (drawing by Nicolas Thomas). (5) Strasbourg (France), Marais Vert, fourteenth century (drawing by Nicolas Thomas). (6) Basel (Switzerland), Ringelhof, mid-thirteenth century (Kamber and Kurzmann 2002). (7) Paris (France), Louvre, mid-fourteenth century (drawing by Stephen Moorhouse in Rouaze 1986). (8–10) Wakefield (UK), Sandal Castle, first half of the fifteenth century (Moorhouse 1983) 59

3.7 Archaeological finds (pottery). (1) Pelican, still-head for cohobation or circulation, Paris (France), Louvre, mid-fourteenth century (drawing by Stephen Moorhouse in Rouaze 1986). (2) Worm, apparatus for coiling the still product, Metz (France), Pontiffroy, fourteenth century (Dautremont et al. 2001). (3) Decorated still-head with two collecting channels for fractional distillation, Orschwiller (France), Haut-Kœnigsbourg castle, unpublished (drawing by Thierry Logel, hypothetical reconstruction by Nicolas Thomas) 60

5.1 The alchemist as an illustration of "Counterfeiters and Deception." Sebastian Brant, *Stultifera navis*, Basel, 1498, fol. 115v. © Bibliothèque nationale de France 98

6.1 Metalworking processes. The central column lists various processes that relate to the primary production of metals and alloys (above) and those carried out during secondary working, turning metals into useable objects (below). To the right are listed the products of each process, which are then the raw material for the next process: for example, mining provides ores, which are the raw material for smelting. Diagram by Justine Bayley 113

6.2 Cupels from the British Isles. Those at the top are early medieval ceramics, while those at the bottom are used (dark) and unused (light) sixteenth-century examples made of bone ash. The scales are in millimeters/centimeters. Top © Justine Bayley; bottom © Nicolas Thomas 117

6.3 A sixteenth-century assay laboratory with furnaces for distilling strong mineral acids (center foreground). Note at the back a pair of crucible furnaces and, top left, an assaying furnace. Lazarus Ercker, *Fleta minor. The Laws of Art and Nature*, 1683, book 1, p. 2. Source: Wellcome Collection 118

6.4 A Bohemian glass furnace of the early fifteenth century CE. Miniature from a manuscript of Sir John Mandeville's travels (BL, MS Add 24189, fol. 16). Source: Alamy 121

8.1 Allegorical figure within the alchemists' flask. Detail of Ripley Scroll, Beinecke Rare Books & Manuscript Library, MS Mellon 41. By permission of the Beinecke Rare Books & Manuscripts Library 151

8.2 Giles Du Wes, exposition of the pseudo-Llullian *Cantilena* (composed ca. 1332), added to the mid-fifteenth-century manuscript ca. 1506. Beinecke Rare Books & Manuscript Library, MS Mellon 12, fol. 163v. By permission of the Beinecke Rare Books & Manuscripts Library 153

8.3 Numbered figures added to the margins of the pseudo-Llullian *Practica Testamenti*. Beinecke Rare Books & Manuscript Library, MS Mellon 12, fol. 119r. By permission of the Beinecke Rare Books & Manuscripts Library 153

8.4 Aristotelian square of opposition in the *Rosarius philosophorum* (incipit "Desiderabile desiderium impreciabile precium") attributed to John Dastin. Beinecke Rare Books & Manuscript Library, MS Mellon 28 (ca. 1525), fol. 4r. By permission of the Beinecke Rare Books & Manuscripts Library 155

8.5 Figures describing a succession of compounds in the pseudo-Llullian *Practica Testamenti*. Beinecke Rare Books & Manuscript Library, MS Mellon 12, fol. 99v. By permission of the Beinecke Rare Books & Manuscripts Library 157

8.6 Alchemical tree in the pseudo-Llullian *Liber de secretis naturae seu quinta essentia*. Beinecke Rare Books & Manuscript Library, MS Mellon 12, fol. 277v. By permission of the Beinecke Rare Books & Manuscripts Library 158

8.7 Circular figures in the pseudo-Llullian *Liber de secretis naturae seu quinta essentia*. Science History Institute, Othmer Library MS 7 (1498). By permission of the Science History Institute 159

8.8 A later reader's attempt to sketch the geometrical figure described in a copy of the *Rosarius philosophorum* (incipit "Desiderabile desiderium") attributed to John Dastin. Cambridge, Trinity College Library MS O.2.18 (late fourteenth/early fifteenth centuries), fol. 88v. By permission of the Master and Fellows of Trinity College, Cambridge 160

8.9 The Emerald Tablet, *Aurora consurgens*. Zurich Zentralbibliothek MS Rh. 172 (fifteenth century), fol. 3r. By permission of Zurich Zentralbibliothek 164

8.10 The coronation of the Virgin, *Das Buch der heiligen Dreifaltigkeit*. John Rylands Library, MS German 1 (fifteenth century), fol. 9r. Copyright of The University of Manchester 166

8.11 Calcination of mercury, *Das Buch der heiligen Dreifaltigkeit*. Leiden University Library, MS Vossianus chym. 29 (sixteenth century), fol. 76r 167

8.12 Threefold mercury, *Aurora consurgens*. Leiden University Library, MS Vossianus chym. 29, fol. 76r 169

8.13 The basilisk repelled by a mirror, *Aurora consurgens*. Zurich Zentralbibliothek MS Rh. 172, fol. 41v. By permission of Zurich Zentralbibliothek 169

8.14 The red elixir, *Donum dei*. Beinecke Rare Books & Manuscript Library, MS Mellon 71 (late seventeenth century), p. 93. By permission of the Beinecke Rare Books & Manuscripts Library 170

8.15 Figure from *Das Buch der heiligen Dreifaltigkeit*. Science History Institute, Othmer MS 10b. By permission of the Science History Institute 171

8.16 The book of seven seals. Ripley Scroll, Princeton University Library MS 93 (late sixteenth century). By permission of the Princeton University Library 172

8.17 The serpent of Eden. Ripley Scroll, Princeton University Library MS 93 (late sixteenth century). By permission of the Princeton University Library 174

8.18 Failed attempt to plot a pseudo-Llullian figure. Beinecke Rare Books & Manuscript Library, MS Mellon 12, fol. 272v. By permission of the Beinecke Rare Books & Manuscripts Library 176

8.19 Decapitation of Sol, *Splendor solis*. Beinecke Rare Books & Manuscript Library, MS Mellon 86, p. 289. By permission of the Beinecke Rare Books & Manuscripts Library 177

8.20 The risen sun, *Splendor solis*. British Library, Harley MS 3469 (1582), fol. 33v 178

LIST OF TABLES

6.1 Some chemical processes and examples of their use 109

6.2 Summary of the colorants and opacifiers found in medieval glass and enamel 124

SERIES PREFACE

A Cultural History of Chemistry examines the history of chemistry and its wider contexts from antiquity to the present. The series consists of six chronologically defined volumes, each volume comprising nine essays; these fifty-four contributions were written and/or edited by a total of fifty scholars, of ten different nationalities. Of Bloomsbury's many six-volume *Cultural Histories* currently in print, this is the first in the physical or natural sciences; it is also the first multivolume history of chemistry to appear since James Riddick Partington's four-volume *History of Chemistry* concluded more than fifty years ago. It is distinguished, among other qualities, by its endeavor to take the subject from antiquity right to the present day.

This is not a conventional history of chemistry, but a first attempt at creating a cultural history of the science. All cultures, including the various branches of natural science, consist of mixed constructs of social, intellectual, and material elements; however, the cultural-historical study of chemistry is still in an early stage of development. We hope that the accounts presented in these volumes will prove useful for students and scholars interested in the subject, and a starting point for those who are striving to create a more fully developed cultural history of chemistry.

Each volume has the same structure: starting with an interpretive overview by the volume editor(s), the eight succeeding chapters explore for each respective era in chemistry its theory and concepts; practice and experiment; laboratories and technology; culture and science; society and environment; trade and industry; learning and institutions; and art and representation. Readers therefore have the option to read multiple chapters in a single volume, thus learning about the cultures of chemistry in a single era; or they may prefer instead to read corresponding chapters across multiple volumes, learning about (e.g.) the art and representations of chemistry through the ages. Though the scope is global, major emphasis is placed on the Western tradition of science and its contexts.

Whether read synchronically or diachronically, in any multiauthor undertaking like this one readers will inevitably notice overlaps and repetitions, conflicting historical interpretations, and (despite the magnitude of the project) occasional gaps in coverage. These are inescapable consequences, but they actually offer advantages to the reader, both in making each chapter closer to self-contained and in demonstrating the dynamism of the discipline; like science itself, the study of its history is ever contested and incomplete.

Chemistry has been called the "central science," due to its fundamental importance to all the other physical and natural sciences. It is the archetypical science of materials and material productivity, and as such it has always been deeply embedded in human industry, society, arts, and culture, as these volumes richly attest. The editors and authors hope that *A Cultural History of Chemistry* will be of great interest and enjoyment not just to chemists and specialist historians of science, but also to social, economic, intellectual, and cultural historians, as well as to other interested readers.

Peter J.T. Morris and Alan J. Rocke
London (UK) and Cleveland (USA)

Introduction

SÉBASTIEN MOUREAU

CHEMISTRY AND ALCHEMY

Writing a cultural history of chemistry in the Middle Ages requires great caution[1]. At that time, chemistry as we know it, namely "the branch of science concerned with the substances of which matter is composed, the investigation of their properties and reactions, and the use of such reactions to form new substances" (*Oxford English Dictionary*, *s.v.* chemistry), did not yet exist as a scientific discipline. It is therefore necessary first of all to adapt our modern understanding to clearly define the concepts that will be used in this book.

In its modern sense, chemistry studies the components of material bodies, their properties and reactions. It differs from physics, which is "concerned with the nature and properties of matter and energy; the subject matter of physics includes mechanics, heat, light and other radiation, sound, electricity, magnetism, and the structure of atoms" (*Oxford English Dictionary*, *s.v.* physics).

What the natural philosophers of the Middle Ages called chemistry is different from our modern chemistry; its meaning is broader. Medieval chemistry was the science of matter, its composition and changes. Thus, medieval chemistry encompassed not only phenomena that we call chemical, such as the dissolution of salt in water, but also phenomena that we call physical, such as the passage of water from the liquid to the solid state. This medieval science is what we now call alchemy. Many terms were used to designate it, especially in the alchemical texts themselves, but it was most often called in the Middle Ages by nonalchemist scholars *kīmiyā'* in Arabic and *alchimia* or *chimia* in Latin (the Latin term being the transcription of the Arabic word with the article *al*). But it must be stressed that, while a clear distinction is made today between chemistry and alchemy, this was not the case in the Middle Ages: *alchimia* and *chimia* have the same meaning in medieval texts (Newman and Principe 1998; Principe and Newman 2001).

So, in what sense should "chemistry" be understood in the title of this volume? It would have been anachronistic to understand the term in the strictly modern sense: this volume will therefore above all present a cultural history of alchemy in the Middle Ages. However, due to the very structure of the series *A Cultural History of Chemistry*, some concepts related to the modern meaning of chemistry are also discussed: for example, industry and crafts will be studied (see Chapter 6), whereas they do not fall directly under alchemy during the Middle Ages.

It is therefore necessary to define alchemy more precisely. In its narrowest sense, alchemy is the attempt to transmute a base metal such as lead or copper into a noble metal such as silver or gold. But even before the Middle Ages, alchemy already had a much broader meaning. From the beginning, alchemy also focused on metal dyeing in general (i.e. metal coloring), gemstone manufacture, and other craft techniques. In the Middle Ages, both in the Arabo-Muslim world and in the Latin West, and even if it was considered only as a simple mechanical art by some (i.e. a simple practice that did not deserve theoretical investigation; Mandosio 1993), alchemy was often approached as a branch of natural philosophy, of science that studies the world. The subjects that it covered are far more numerous than just transmutation: from surface alloys to the coloring of glass, from philosophical reflections on the material and its transformations to descriptions of the furnaces (for medieval definitions of alchemy in both Arabic and Latin cultures, see Moureau 2020). Certain fields that are sometimes far removed from alchemy in the strict sense have been developed, such as the creation of the homunculus (i.e. a miniature man) in the Jābirian corpus or the prolongation of life in Roger Bacon's works. And alchemy had links with other sciences, such as mineralogy and metallurgy, from which it is sometimes even difficult to distinguish in some texts. Alchemical treatises also often abound with biological, medical, astronomical, astrological, or other considerations, and even sometimes with gnostic and mystical reflections. Medieval alchemical texts were protean. There were often very different types of content within the texts themselves: theoretical sections, practical parts (often abundant in recipes), or allegorical content. The form of the treatises is also very variable from one text to another: alchemical texts can be written in the form of prose, poetry, dialogues, recipes, commentaries, or doxographies. In addition, an allegorical meaning soon emerged, and works such as the eleventh-century *Alchemy of Happiness* (*Kīmiyā' al-sa'āda*) of Ghazālī were soon written. A religious sense also quickly developed, which is often difficult to address because of the preponderance of religion in any medieval written work.

Alchemy must also be distinguished from art and crafts in their modern sense. In this volume, craft is understood with a specific meaning: it is first and foremost handicraft, the practices of craftsmen and artists, and not only art or fine arts in the modern sense. If alchemy was often quite close to art and craftsmanship in terms of techniques and practices, it was clearly different in terms of theory: while medieval craftsmen and artists generally did not develop a theory to explain their operations, theoretical speculations were present with most alchemists. Moreover, alchemy was a science of the literate, unlike craftsmanship. Alchemical treatises were the products of scholars, whether they were practitioners or not. The aim was also different: craftsmen wanted above all to produce substances or artifacts, while alchemists had different intentions not only in their practices, but also when they wrote their texts. Scientific and theoretical intentions were absent from the craft industry. And the approach was different. However, in practice, many craftsmen's recipes were found in alchemical texts.

Another more prosaic reason for this volume to be devoted above all to alchemy is that, unlike medieval craftsmanship, alchemy has not been studied very much to date. Indeed, although it has enjoyed a strong resurgence of interest over the past two decades (there is currently an efflorescence in the study of the history of alchemy), publications on this subject remain few and far between, and especially are still not widely read outside the strict domain of the history of alchemy (Principe 2013: 1–4). In particular, the history of Arabic alchemy remains the poor relation of the discipline. And this state of affairs is quite understandable: it is due to the difficulties inherent in the study of the history of alchemy. This requires not only very specific philological and linguistic skills, but also extensive historical and chemical knowledge. Few researchers are trained to work on this subject. However, the renewal of studies on the history of alchemy in recent decades has made it possible to carry out the first syntheses (Newman 2013; Principe 2013; Calvet 2018a; Rampling 2020). But to date, there is no synthetic appraisal on the history of alchemy in both the Arabo-Muslim world and the Latin West. The purpose of this book is to present a first study dealing with the two civilizations, Arabo-Muslim and Latin-Christian, in parallel. In this volume, the reader will find a presentation of current knowledge on the history of alchemy and can go further if she or he wishes thanks to the references proposed in the notes. Since most chapters will deal with the Arabo-Muslim world and the Latin world, dual dates and dual centuries often appear. They correspond to the two eras used: AH refers to the years of the Muslim calendar, the beginning of which is the Hegira, and BCE and CE refer to the Christian calendar. Some authors in the volume use the word Islamicate, which is the term now commonly used to refer to the cultures within the Islamic realm, which (unlike the terms Arab, Arabic, Muslim, or Arab-Muslim) avoids identifying these cultures either with a language, ethnic group, or religion. Chinese and Indian alchemy are not presented in this volume, as they do not actually correspond strictly speaking to the definition of alchemy as it developed in the Arabo-Muslim and Latin worlds. Moreover, they have been even less studied, and the sources are for the most part inaccessible (on Chinese alchemy, see the preliminary studies Sivin 1968; Sivin 1990; on Indian alchemy, see the preliminary studies Ray 1956; Wujastyk 1984; Hellwig 2009).

This observation makes it possible to address here the main difficulties related to the study of medieval alchemy. The current state of research in this field, although not in its infancy, is, as it has been said, not highly advanced. This is particularly true for access to sources. There are many alchemical sources, both Arabic and Latin, but most of them are still inaccessible to the nonspecialist: most of the texts are unpublished, and the use of manuscripts is necessary when working on the history of alchemy. This dire lack of publications is particularly true for Arabic alchemy, of which several fundamental texts have not yet received a critical edition. It should also be added that, although there are many medieval Arabic and Latin texts, the ancient Greek alchemical texts that are at the origin of alchemy have been preserved in very small numbers, although more often already published (on Greek antique alchemy, see Berthelot and Ruelle 1887–8; Letrouit 1995; Martelli 2013). Syriac alchemical texts, which in all likelihood were one of the links between ancient Greek alchemy and medieval Arabic alchemy, are hardly studied at all (on Syriac alchemy, see Berthelot et al. 1893: vol. 2; Martelli 2013; Martelli 2014a).

After this first difficulty, other problems appear to the researcher who today works on a medieval alchemical text. Several characteristics inherent in the alchemical genre stand in the way of the reader. The first is the strong predilection of alchemistic authors for pseudepigraphy (i.e. ascribing false names of authors to texts). It is indeed very common for an alchemical text to be attributed, whether by its actual author or by posterity, to another author, generally an authority of the time. Thus, prominent figures such as Aristotle, Plato, Apollonios of Tyana, Avicenna, Thomas Aquinas, and others had alchemical texts attributed to them. In addition to these, there was a plethora of mythical and legendary authors such as Hermes, Agathodaimon, Chymes, etc. There were many reasons for this propensity for pseudepigraphy, including the desire to give the text a wider audience, which was probably one of the main reasons. This multiplicity of apocrypha gives rise to a few difficulties for the modern researcher. To make things even more complicated, a text is often attributed to different authors in different manuscripts. And this is without counting on the variability of book titles as well; alchemical texts are often known under different titles, and several treatises also have identical titles, hence the use of sometimes naming them by citing their incipit (i.e. the first words of the text). These complex issues are often confusing, as the reader will see in this volume. It is sometimes difficult to find one's way around: the concepts of author and alleged author must be clearly distinguished, and a text is sometimes not solidary to its title. The reader of this volume should therefore be careful when reading the name of an alchemist (see below for an example with Jābir b. Ḥayyān). Another major problem, which is more technical, is the great fluidity of alchemical traditions. Like many scientific textual traditions, the alchemical tradition was a living tradition. The texts evolved over time, and the multiple scribes were often more than copyists, intervening in the texts to add or delete elements. For this reason, editing an alchemical text is most of the time a challenge.

In addition to the form, the study of alchemy is also problematic in its content. Indeed, one of the well-known characteristics of alchemical texts is the abundance of coded terms. Reading an alchemical treatise can often be frustrating, as many of the words are coded and the consistency of these codes is rarely ensured. While some terms are traditional and easily identifiable, such as the "fugitive" (*farrār* in Arabic, or *fugitivus* in Latin) to designate mercury, most are difficult to decipher. Moreover, these codes have themselves been interpreted from text to text, changing meanings frequently. And even when they are not coded, the substances are not always the same in different places and at different times: what is called sulfur in Baghdad in the tenth century may be different from what is called sulfur in al-Andalus (Iberian Peninsula under Muslim rule) in the thirteenth century, and not correspond to what is called sulfur today. And the terms used to designate substances in the Middle Ages usually have a broader meaning than today. For instance, for many alchemists, there are several different golds, depending on their purity, origin, and composition, which is not the same as the conception of a modern chemist. Thus, the practices described in the texts are often inaccessible to the modern reader. Moreover, even when substances can be identified with certainty, recipes remain difficult to comprehend because of the impurity of the substances used in the past. Medieval alchemists used impure ingredients, which have various effects in chemical and physical reactions. Sometimes impurities are even the key elements of recipes, which do not work if the pure ingredients are used (Principe 1987).

The volume follows the same structure as the other volumes in the series. Each chapter deals with both the Arabo-Muslim civilization and the Latin world. Most of the chapters in this volume are written by several authors. For Chapters 4–7, the main contributors are joined by secondary contributors, whose names are indicated after the preposition "with" at the beginning of the chapters, as well as under the titles of the sections they have written.

The remainder of this introduction is devoted to a brief history of alchemy in the Middle Ages, which will allow the reader to place the information contained in the rest of the volume in a chronological context.

A BRIEF HISTORY OF ALCHEMY IN THE MIDDLE AGES

The history of alchemy is usually divided into three main periods: Egyptian–Hellenic alchemy, Arabic alchemy, and Latin alchemy. In this book, we look at alchemy in the Middle Ages, so we work on early Arabic alchemy, which mainly inherited its concepts from late ancient Greek alchemy, and on the beginnings of Latin alchemy. From a chronological point of view, the studied period goes from the reception of the Greek heritage in Arabo-Muslim civilization around the eighth century to the advent of Paracelsism (this not being included) in the Latin world (i.e. the end of the fifteenth century). In each chapter, the two civilizations, Arabic and Latin, are discussed, often but not exclusively by different specialists.

Reception of Greco-Egyptian alchemy

It is not possible to date the birth of alchemy. The sources that have reached us are very incomplete for the older periods, and it is not possible to define precisely what preceded the appearance of the first preserved sources. The oldest alchemical sources that have been preserved date back to the first centuries of the Christian era in the Egyptian–Hellenic world.

Greek alchemy was transmitted to the Arabo-Muslim world during the movement of translations from Greek into Arabic from the mid-eighth century to the end of the tenth century (Gutas 1998; Martelli 2017). It also passed through Syriac intermediaries, and a corpus of alchemical texts in this language is preserved (Martelli 2014a; Takahashi 2015). Arabo-Muslim scholars, when receiving alchemy, assimilated it very quickly, and major works were composed as early as the ninth or tenth century. It should be noted, however, that, while it is undoubtedly possible to affirm this passage of alchemy from Greek to Arabic, there are very few Arabic alchemical texts of which the Greek original is known to us. The main reason for this is most probably the lack of preserved texts from the Greek corpus. The Greek alchemical texts that we have preserved are found mainly in quotations in the Arabic alchemical texts, in which they abound.

The close links between Greek and Arabic alchemy can be observed on several levels. First of all, the vocabulary is largely borrowed from Greek, at least for the theoretical part of alchemy; the more practical part, in particular that related to craftsmanship, is often borrowed from Persian. For instance, the elixir, *iksīr* in Arabic, comes from the Greek *xērion*, and has a Syriac counterpart *ksirīn*; alchemy itself, *kīmiyā'* in Arabic, comes from the Greek *chēmeía*, in Syriac *kimiyo*. Even more than vocabulary, Arabic alchemical doctrines are generally, in the oldest texts, directly those of Greek alchemists, assimilated and developed. For example, the Aristotelian theory of double exhalation is at the root of the theory of metallogenesis in Arabic texts, and the Galenic theory of the four intensities of elementary properties also appears in Arabic alchemy. But an indisputable mark of Greek influence on Arabic alchemy is observed in the authorities cited by Arabic alchemists. They mention a plethora of Greek authors, real or legendary. While the works to which they refer are often apocryphal, they are nevertheless clearly influenced by Greek culture. As for the Persian influence on Arabic alchemy, while it is indisputable in practical vocabulary (e.g. *zi'baq*, mercury, from Persian *žīwah*, for *jīwah*), as has been said, it is not documented and cannot be studied precisely to date.

The first phase of assimilation of Greek alchemy is characterized not only by the translation of Greek texts, but also by the writing of texts in Arabic then attributed to Greek authorities; they are, so to speak, fake translations. For example, the *Tome of Images* (*Muṣḥaf al-ṣuwar*) is an Arabic treatise attributed to Zosimos of Panopolis but actually composed in Arabic on the basis of a translation of an authentic writing by Zosimos of Panopolis; and the *Book of the Fourths* (*Kitāb al-rawābi'*) is an alchemical treatise written in Arabic and ascribed to Plato.

The authors of this period of assimilation are often unknown and there are many pseudepigraphs. Among the great names, Zosimos occupies a prominent place (on the Greek Zosimos, see Letrouit 1995: 22–46; Mertens 1995: i–clxix; on the Arabic Zosimos, see Abt 2007a; Abt 2007b; Hallum 2008; Hallum 2009; Abt and Fuad 2011). Some of his authentic texts were translated from Greek into Arabic, but others, such as the abovementioned *Muṣḥaf al-ṣuwar*, were apocryphal compositions. The Greek authorities were both mythical, such as Hermes, Agathodaimon, and Ostanes (a Persian magician cited in Hellenistic Greece who may have come from a Babylonian background), and historical, such as Pythagoras, Archelaos, Socrates, Democritus, Plato, and Aristotle. Some of these are just famous names not related to identified alchemical texts, whether authentic or pseudepigraphic in Greek or Arabic, such as Pythagoras and Socrates; others refer to translated Greek pseudepigraphic texts (perhaps only in the form of quotations), such as Democritus; still others are directly used for pseudepigraphs written in Arabic, such as Aristotle, Plato, or Archelaos. The alchemical authors also "alchemized" certain texts that were not strictly alchemical at first. Thus, the famous *Secret of Creation* (*Sirr al-khalīqa*), falsely attributed to Apollonius of Tyana (Balīnūs), is a treatise on cosmology that was soon after its composition considered an alchemical text by alchemists and commented on as such (Weisser 1979; Weisser 1980; Pappacena 2000; Travaglia 2001).

A legend illustrates how Arabic alchemists put forward the Greek origin of their knowledge. Several Arabic authors claimed that the first Arabic alchemist was the Umayyad Prince Khālid b. Yazīd who lived in the first century AH/seventh century CE. In this story, Prince Khālid b. Yazīd, wanting to know the secrets of alchemy, learned that a Byzantine (*rūmī*) hermit monk named

Maryānus was in possession of them. Khālid asked for him, and, after days and days of discussions, Maryānus told him his secrets (Ruska 1924; Sezgin 1971: 120–6; al-Hassan 2004; Bacchi and Martelli 2009; Dapsens 2016; Dapsens forthcoming). Manfred Ullmann showed in 1978 that this legend was actually constructed from pseudepigraphic literature dating back to the ninth century at the earliest. But the references of the alchemists were not only oriented toward Greece. Among the alchemical authorities, there are also several mythical figures who were already part of the Muslim tradition, such as Adam, Noah, or Korah.

Arabic alchemy

After the period of assimilation, a properly Arabic alchemy arose. Actually, this Arabic alchemy appeared at a very early stage, since some of the most important alchemical Arabic writings date from the ninth century onwards. A few synthetic studies have been devoted to Arabic alchemy, although, as has been said earlier, this field of research remains, if not completely a *terra incognita*, at least a *terra non bene nota* (Sezgin 1971; Ullmann 1972: 145–270; Ullmann 1986; Anawati 1996; Forster 2016).

As mentioned above, the author most commonly considered as the first Arabic alchemist by the tradition is Prince Khālid b. Yazīd. However, other Arabic authorities were not forgotten, and alchemical texts were also attributed to the first-century AH/seventh-century CE caliph ʿAlī b. Abī Ṭālib, or to the second-century AH/eighth-century CE imām Jaʿfar al-Ṣādiq, although their legends did not reach similar proportions to the one on Khālid.

The most famous Arabic alchemist is undoubtedly Jābir b. Ḥayyān. To this character, probably legendary, is attributed a corpus of nearly 3,000 texts (a number that must be put into perspective, however, since many of these texts have only reached us by their title, and many of those we have preserved are only a few pages long). According to tradition, Jābir b. Ḥayyān lived ca. 107–193 AH/725–812 CE, and he was the disciple of the Shīʿite imām Jaʿfar al-Ṣādiq. The texts in the corpus are often of Shīʿite tendency, and focus mainly but not only on alchemy: the other sciences are also treated in multiple places. However, since Paul Kraus's masterful study on the subject (Kraus 1942; Kraus 1943), it has been shown that the corpus cannot have been written by one and the same person; Kraus went so far as to argue that the character of Jābir b. Ḥayyān himself is a legend created *a posteriori*. Most researchers now accept this hypothesis (Ullmann 1972: 198–208; Delva 2017), while others have suggested the possibility of a historical Jābir, although in a much less convincing way (Sezgin 1971: 132–269; Haq 1994: 8–32). Whatever may be the historicity of Jābir b. Ḥayyān as person and author, the Jābirian writings had an influence on all subsequent Arabic alchemical texts. The reader should therefore be careful when reading in this volume a mention of texts from Jābir b. Ḥayyān: this is the so-called corpus Jābirian, texts that have been attributed to this author and have come down to us under this attribution. What is more, the Latin Geber is also a complex matter (see below).

Another major figure in Arabic alchemy, the authenticity of whose works is not disputed this time, is the philosopher, physician, and scientist Abū Bakr al-Rāzī of the second to third centuries AH/ninth to tenth centuries CE, called Rhazes by the Latins (Sezgin 1971: 275–82; Ullmann 1972: 210–13). His alchemy is characterized by a very well-developed practical side and a significant empirical component.

Other trends of alchemy are found in Arabic alchemy. Several allegorical alchemical writings were written by Muḥammad b. Umayl al-Tamīmī, an alchemist who lived in Egypt in the first half of the fourth century AH/ tenth century CE (Sezgin 1971: 283–8; Ullmann 1972: 217–20). In his *The Silvery Water and Starry Earth* (*Kitāb al-mā' al-waraqī wa-al-arḍ al-najmiyya*), he commented on his own allegorical alchemical poem entitled *Epistle of the Sun to the Crescent Moon* (*Risālat al-shams ilā al-hilāl*; edited in Stapleton et al. 1933).

Alchemy also reached the Western part of the Arabo-Muslim world. The first Arabic alchemical text composed in al-Andalus that is known to us today is the *Rank of the Sage* (*Rutbat al-ḥakīm*). This text, which, according to the most likely hypothesis to date, was written by the specialist in Muslim tradition (*muḥaddith*) Maslama b. Qāsim al-Qurṭubī between 339/950 and 342/953 (Fierro 1996), is the older sister of a much better-known text, the *Aim of the Sage* (*Ghāyat al-ḥakīm*), of which the Latin translation entitled *Picatrix* was one of the most read astral magic texts in the medieval Latin world. In the *Rutba*, Maslama imports into Andalus a large number of sources and doctrines that open the way to a Western Arabic alchemy (Holmyard 1924; Carusi 2005; Callataÿ 2013; Madelung 2014–2015; Callataÿ and Moureau 2015; Callataÿ and Moureau 2016; Callataÿ and Moureau 2017; Madelung 2017a; Madelung 2017b). Another great figure, but this time outside al-Andalus, in Fez, Morocco, is the preacher and alchemist Ibn Arfa' Ra's, who died in 593 AH/1197 CE (Ullmann 1972: 231–2). This alchemist composed a collection of forty-three alchemical poems entitled the *Gold Nuggets* (*Shudhūr al-dhahab*), which was a best seller of the time (Todd 2016).

In the fifth- to sixth-century AH/eleventh- to twelfth-century CE East, al-Ṭughrā'ī is often considered one of the greatest authorities in the field by alchemists. He was a senior official of the Seljuk Malik Shāh and then Muḥammad I, then executed by Maḥmūd III after the revolt of his brother Mas'ūd b. Muḥammad. In addition to his poetic literary work, al-Ṭughrā'ī wrote several alchemical works in prose and poetry that achieved great success (Kouhkan 2015). A century later, in the middle of the seventh AH/thirteenth CE century, another great figure in alchemical literature appeared, Abū al-Qāsim al-'Irāqī, also called al-Sīmāwī, whose life remains almost entirely unknown to us (Holmyard

1926; Ullmann 1972: 235–7). Besides an important doctrinal work, the *Book of Knowledge Acquired Concerning the Cultivation of Gold* (*Kitāb al-ʿilm al-muktasab fī zirāʿat al-dhahab*; edited in Holmyard 1923), he wrote a highly original text presenting an alchemical interpretation of Egyptian hieroglyphics, the *Seven Climes* (*Al-aqālīm al-sabʿa*); the manuscripts of this text are mostly decorated with mysterious illuminations. Al-Sīmāwī was also versed in magic (Saif 2016).

In the eighth century AH/fourteenth century CE, the alchemist al-Jildakī distinguished himself by the quality and quantity of his writings (Holmyard 1937; Harris 2017). This author, whose life is little known to us, wrote a very large corpus of texts that were widely disseminated, as shown by the impressive number of manuscripts of his preserved works. He has often been considered a simple compiler, but this *a priori* is probably due to the fact that he has been studied very little to date and that a goodly number of his works are very little known.

The transmission of Arabic alchemy to the Latin West

During the twelfth and thirteenth centuries, a large movement of translations from Arabic into Latin took place in the areas of contact between the Arabo-Muslim and Latin worlds, mainly the Iberian Peninsula and Italy. Although this transfer of culture had already begun during the eleventh century, it was above all during the following two centuries that it was amplified (Burnett 2009; Burnett 2013). In this period, a flow of new knowledge reached Latin scholars through translators. Alchemy stands out among this new knowledge, as it was previously unknown in the Latin West, with the exception of a few recipes that circulated in books on other practices (on the translation of alchemical texts, see Moureau 2020).

In the Latin West, the context of arrival of these translations is above all composed of books of technical craft recipes such as, around 800, the *Compositiones variae* (Johnson 1939) or, in the ninth century or earlier, the *Mappae clavicula* (edited and translated in Smith and Hawthorne 1974). One exception should be mentioned, however, in the presence of a recipe for "Spanish gold" in the *Diversarum artium schedula* of the monk Theophilus (perhaps Roger of Helmarshausen; edited and translated several times, see L'Escalopier 1843; Ilg 1874; Hawthorne and Smith 1979; Dodwell 1986).

The translators who worked on alchemical texts are generally unknown, with the exception of a few (Robert of Chester, Gerard of Cremona, Michael Scot). The methods used by these translators for alchemical texts were similar to those used for other disciplines, and they were often assisted by Jewish or Muslim scholars who were familiar with the Arabic language (Burnett 2001; Burnett 2002; Burnett 2008). It is difficult to define the motivations that led the translators to take an interest in alchemy (Moureau 2020: 127–8); only one testimony has come down to us, in the preface to the translation of the *Liber de compositione alchemiae*, perhaps by Robert of Chester (Ruska 1924: 35; Kahn 1990–1; Lemay 1990–1). In this preface, the translator explains that he is translating the text in order to introduce the Latin world to a new science. Filling a gap in the curriculum was therefore probably one of the objectives of the translators of alchemical texts, and the theoretical and practical contribution of alchemical texts to the Latin world was far from negligible. In addition to this is probably added the important place that alchemy held in the classifications of sciences in the Arabo-Muslim world, as well as perhaps also the lure of gain represented by the possibility of transmutation. To date, more than forty alchemical texts translated from Arabic into Latin are known, which shows the interest of the translators (Moureau 2020). Among these texts, it is difficult to identify selection criteria. Indeed, the translators dealt with all types of content, from the most theoretical texts to allegories, practical works, and recipe books. Nor is the form distinctive: poetry, prose, dialogues, recipes, doxographies, and commentaries can be found. The doctrines present in the texts are likewise not distinctive, and various alchemical systems can be found in the translations. The only criterion that seems to emerge is the authority of the authors of the texts, whether they were authentic or not. An alchemical text was more likely to be translated if it was attributed to an authoritative figure. Thus, one finds texts attributed to Plato (*Book of the Fourths*, *Kitāb al-rawābiʿ*/*Liber quartorum*), Aristotle (*Book of the Stones*, *Kitāb al-aḥjār*/*Liber lapidum*), Mary the Jew (*Epistle of Mary, Daughter of King Sāba the Copt, to Aros*, *Risālat Māriya bint Sāba al-malik al-qibṭī ilā Āras*/*Practica Mariae prophetissae sororis Moysi*), Avicenna (alchemical *De anima*; *Epistle on the Elixir*, *Risālat al-iksīr*/*Epistola ad Hasen regem*), etc.

The first complete translation of a dated alchemical text is the *Book of the Composition of Alchemy* (*Liber de compositione alchemiae* attributed to Morienus, the translation of the *Masāʾil Khālid li-Maryānus al-rāhib*; Latin text edited in Stavenhagen 1974; recent *status quaestionis* in Dapsens 2016).[2] This work is said to have been translated by Robert of Chester in 1144, and it would be the first complete alchemical work to have penetrated the Latin West. However, this date is more of a symbolic milestone than a precise actual date, since most Arabic–Latin translations of alchemical texts cannot be dated. The main Arabic alchemical trends can be found in early Latin alchemy, and the translations ensured the transmission of a corpus large enough to convey an understandable doctrine. The alchemy of Arabic–Latin translations is dominated by the Jābirian concept of elixir (see Chapter 1). Out of the large number of Jābirian works written in Arabic, Latin translations were made of the *Book of the Seventy* (*Liber de Septuaginta*, translation of the *Kitāb al-sabʿīn*; incomplete edition on the basis of only one manuscript in Berthelot 1906: 310–63; edition of the third treatise of the work in Colinet 2000a: 179–87), the *Book of Mercy* (*Liber misericordiae*, translation of the *Kitāb al-raḥma*; edited in Berthelot et al.

1893: 3: 132–60 (ar. pp.); French translation in Berthelot et al. 1893: 3: 163–90), and the *Book of Kingship* (*Liber regni*, translation of the *Kitāb al-mulk*; edited in Newman 1994b: 288–93); they circulated under the name of Geber (i.e. the Latinized name of Jābir b. Ḥayyān). Other texts, but that were written directly in Latin, circulated under the name of Geber (i.e. were attributed in Latin to an Arabic author and presented as false translations [see below]). Other works convey the Jābirian concepts without necessarily being directly linked to Jābir. For instance, the pseudo-Avicennian *De anima*, a work that is the compilation and translation of three Arabic treatises, perhaps translated in 1226 or 1235, spread widely in the Latin world the idea of making a Jābirian elixir by means of organic substances (studied, edited, and translated into French in Moureau 2016a; Moureau 2016b). The more technical alchemy of Rāzī also appears among the translations, especially the *Book of Secrets of Bubacar* (*Liber de secretorum Bubacaris*, a translation of Rāzī's *Kitāb al-asrār*; Ruska 1935). Other Arabic–Latin translations of a more technical nature gave this very practical alchemy a wide diffusion, such as the treatise *On Alums and Salts* (*De aluminibus et salibus*), an Arabic technical work dating from the eleventh or twelfth century whose author is unknown, which was translated into Latin and of which there is also a Hebrew version (Colinet 2000b: xlii–xlv; Ferrario 2004; Ferrario 2007).[3] Ibn Umayl's allegorical alchemy also passed to the West in the translations of the *Epistle of the Sun to the Crescent Moon* (*Epistola solis ad lunam crescentem*, translation of the *Risālat al-shams ilā al-hilāl*) and of the treatise *On Alchemy* (*De chemia*, translation of the *Silvery Water and Starry Earth*, *Kitāb al-mā' al-waraqī wa-al-arḍ al-najmiyya*). Another vector of Arabic alchemical ideas in the Latin world is the *Turba philosophorum* (a translation of an Arabic text), a doxographic collection written in the form of a dialogue between nine pre-Socratic philosophers including Pythagoras, whose diffusion was incredibly wide (Arabic fragments edited in Ruska 1931; edition and French translation of the Latin text in Lacaze 2018; see also Plessner 1975; Hallum 2009; Kahn 2010b).

The penetration of Arabic alchemy into the Latin West offers a rare case study for the modern researcher. It allows one to study how a completely new science was translated, assimilated, and then developed in a civilization. The passage of alchemy from the Arabo-Muslim world to the Latin West took place without a significant change in the concepts and doctrines, and the fine understanding of the Latin scholars in dealing with occasionally confused and obscure translations sometimes leaves the modern reader filled with admiration.

Early Latin alchemy

The assimilation phase of alchemy was fairly rapid. Indeed, Latin compositions appeared as early as the thirteenth century (for recent studies on early Latin alchemy, see Principe 2013; Calvet 2018a). In the beginnings of Latin alchemy, the concepts of Arabic alchemy were for the most part retained, with some minor modifications. Alchemy had already penetrated the Latin scholarly world sufficiently to be, as early as the middle of the thirteenth century, an important subject in the *Speculum naturale* and *Speculum doctrinale* of Vincent of Beauvais. In these works, this Dominican encyclopedist described alchemy to the scholars of his time, drawing on a series of important alchemical texts that were current in his day (Moureau 2012). Theologians also debated it, such as Albert the Great in the thirteenth century, who became interested in alchemical texts and was the first in the Latin West to establish a mineralogy by writing his *De mineralibus* (edited in Borgnet 1890). His curiosity about alchemy favored the attribution to him of a series of alchemical texts that were in all likelihood apocryphal, such as the *Semita recta* and the *Calistenus* (Kibre 1942; Halleux 1982; Calvet 2012). Thomas Aquinas also addressed the subject, but without the same interest as Albert the Great; yet, he considered transmutation to be possible, although hard to achieve (Crisciani 2006). Certain alchemical texts have been attributed to him, among them the famous *Aurora consurgens*, a text on whose alchemical imagery much ink has been spilled (edited in Franz 1957; on the posterity of Thomas's opinion on alchemy and the relations between scholastics and alchemy, see Matton 2009). Other scholars opposed alchemy, like Giles of Rome in the thirteenth to early fourteenth centuries (Newman 2004: 53–4). In 1317, Pope John XXII issued the bull *Spondent quas non exhibent*, which was directed against fraudsters and alchemists who counterfeited gold (Halleux 1979: 124–6). More vehemently, the general inquisitor of Aragon, Nicolas Eymerich, published a *Contra alchimistas* (1396) in which he condemned alchemists: Nicolas denied the possibility of transmutation for man and asserted that disappointed alchemists had recourse to demons to perform transmutation (Newman 2004: 91–4). The text that was in all likelihood one of the main causes of this vivid debate over the possibility of transmutation was the *Sciant artifices*. This short text is a passage from *fann* 5 of the *ṭabīʿiyyāt* (*Physics*) of Avicenna's *Book of Healing* (*Kitāb al-Shifāʾ*); it was appended by Alfred of Shareshill to his translation of Book IV of Aristotle's *Meteorology*, without mentioning the real author, and therefore considered an authentic work of Aristotle during the Middle Ages. In this passage, Avicenna (considered to be Aristotle) denied the possibility of transmutation (see Chapter 4); the attribution to Aristotle gave considerable weight to the argument (Mandosio and Di Martino 2006). Therefore, alchemy was a frequent topic of scholastic *quaestiones* (Matton 2009). Yet, although it was debated vigorously in the scholarly world, it always remained on the margins of the university curriculum, mainly because it was considered by many scholars to be a mechanical art (Mandosio 1993; see Chapter 4).

One of the first scholars to write alchemical treatises in Latin, in the late twelfth and early thriteenth centuries, was Michael Scot. Active initially in

Toledo and later at the court of Frederick II of Hohenstaufen in Sicily, Michael Scot is credited to have translated from Arabic and written original alchemical texts (Brown 1897; Thomson 1938; Fauré 2006). Various texts are attributed to him, including the translation of a *Lumen luminum*, also called *Liber Dedali* (edited in Brown 1897: 240–69), and the composition of the *Ars alchemiae* (edited in Thomson 1938). The doctrine read in these texts is strongly influenced by Arabic alchemy.

The Franciscan Roger Bacon is one of the pivotal figures of thirteenth-century alchemy. In addition to giving alchemy a prominent place in his doctrine, he was the first in the West to develop medical alchemy. He proposed to use alchemy and its techniques not only to make usual medicines prescribed by physicians, but also to elaborate a properly alchemical preparation that could heal the body and prolong life to the term set by God (Newman 1994a; Newman 1995; Newman 1997; Paravicini Bagliani 2003; Matus 2013). This idea of *prolongatio vitae* met with great success in the West; medical alchemy is one of the characteristics that distinguishes Latin alchemy from Arabic alchemy, in which medical alchemy has never been of central importance. In addition to his authentic works, there are numerous apocryphal texts attributed to Roger Bacon.

At the end of the thirteenth century a work was composed that became a fundamental document of alchemical doctrine until the seventeenth century: the *Summa perfectionis* (edited and translated into English in Newman 1991). This text was possibly written by Paul of Taranto, and pseudonymously attributed to Geber (the Latin name of Jābir b. Ḥayyān), and it circulated as a fake translation from Arabic. The *Summa perfectionis* propagated a theory according to which the elixir used for transmutation can be elaborated from mercury alone, thus opposing the theory (notably Jābirian) of organic elixirs. Other texts written in Latin were attributed to Geber, forming with the *Summa perfectionis* what is called the pseudo-Geberian corpus, a series of fake translations ascribed to one of the most famous names in alchemy.[4] This corpus does not exhibit theoretical unity but was considered the work of a single author by later alchemists.

To the famous Catalan physician Arnald of Villanova, who lived in the late thirteenth and early fourteenth centuries, was ascribed an apocryphal series of twenty alchemical texts (Calvet 2011). These texts were mainly influenced by the alchemy of the *Summa perfectionis*, the pseudo-Avicennian *De anima*, and works of Roger Bacon. The most famous text is the *Rosarius philosophorum* (incipit: *Iste namque liber*), a treatise whose systematization is reminiscent of that of the *Summa perfectionis*. The works attributed to Arnald of Villanova are characterized by a pronounced presence of Christian ideas, going as far as offering a comparison between mercury and Christ. This corpus also contains several treatises on medical alchemy, which contributed to the enthusiasm for potable gold (*aurum potabile*).

Another great scholar of the same period, Ramon Llull, was soon after his death in 1315/1316 regarded as the author of a collection of apocryphal alchemical texts (Pereira 1989). The works that constitute this corpus were not written by the same hand, and the influences that run through it are multiple. The major work is certainly the *Testamentum*, a text that presents an encompassing doctrine of alchemy concerning metals and stones as well as the human body, and whose posterity was of capital importance (edited in Pereira and Spaggiari 1999). Along with the *Summa perfectionis* and the pseudo-Arnaldian corpus, the pseudo-Llullian collection is one of the pillars of medieval alchemical doctrine.

In line with the development of alcohol distillation in the fourteenth century, the Franciscan John of Rupescissa (Jean de la Roquetaillade) elaborated an alchemical system based thereon and assimilated the concept of quintessence to it (Halleux 1981). In his main work, the *De consideratione quintae essentiae* (a best seller at the time, judging from its particularly abundant manuscript tradition), he used the concept of quintessence, which he considered to be the counterpart of the superlunary ether in the sublunary world, and identified it with *aqua ardens* (distilled alcohol): his alchemical system was therefore based above all on repeated distillations. John of Rupescissa is one of the major authors of medical alchemy, and his recipes aimed not only at transmutation, but above all at prolonging life.

In his *Pretiosa margarita novella* written around 1323–1330, the physician Petrus Bonus (Pietro Bono) of Ferrara described alchemy not as a simple mechanical art but as a science (Crisciani 1973). Moreover, the author, who was well versed in philosophy, expressed in his text, written with a scholastic approach, the idea that alchemy is a supernatural and divine science, inaccessible without revelation.

English soil also produced many alchemists (Rampling 2020), such as, in the thirteenth century, Walter Odington (also known as Walter of Evesham), author of the *Icocedron* (Thomas 1971), or, in the fourteenth century, John Dastin, to whom several alchemical texts are attributed (Theissen 1986; Theissen 1991; Theissen 1999; Theissen 2008; Rodríguez Guerrero 2010–13: 92–101), and Philip Oliphant (Beaujouan and Cattin 1981). A century later, England witnessed the birth of one of the gems of alchemical poetry: the *Compound of Alchymie ... Divided into Twelve Gates*, a treatise of pseudo-Llullian inspiration attributed to George Ripley (Rampling 2010; Rampling 2012).

The fifteenth century saw the production of alchemical manuscripts increase considerably. At that time, alchemy was also translated and sometimes composed in vernacular languages. This century was also, together with the sixteenth century, the century of an intense development of alchemical imagery that is well known today (Obrist 1982).

CONCLUSION

In the following pages, the reader will find eight chapters in which she or he will discover first and foremost a cultural history of alchemy in the Middle Ages, with numerous forays into the field of craftsmanship (especially Chapters 2, 3, and 6). The history of chemistry in the modern sense of the word is indeed meaningless for the period studied here, as it is not a discipline in itself at that time.

The Middle Ages run from the fifth to the fifteenth centuries (i.e. over a very long period of time), and the geographical areas covered – the Arabo-Muslim world and the Latin West – are also very broad. The vision proposed in the volume is therefore necessarily very synthetic and all-encompassing. This history must also be read keeping in mind that alchemy is not monolithic, and that research in this field is far from being advanced enough to give a perfect and very clear vision of it, although it is already possible to propose a solid synthesis. With this in mind, the pages of this book abound with references that will allow the reader who so wishes to investigate the subject in greater depth.

Finally, before beginning the reading of the rest of the volume, it is appropriate to abandon the biased vision of the alchemist conveyed in contemporary imagery. Medieval alchemists were not crazy old men chasing chimeras, but scientists of their time, eager to study nature and understand its visible and invisible facets. Alchemy is not a far-fetched science of the Middle Ages, and some of the greatest scholars of the time were interested in it and even devoted themselves to it.

CHAPTER ONE

Theory and Concepts: *The Shared Heritage of Byzantine, Arabo-Muslim, and Latin Alchemy*

MATTEO MARTELLI, SÉBASTIEN MOUREAU,

AND JENNIFER M. RAMPLING

The theories of medieval alchemy are both technically complex and intricately self-referential.[1] Over the space of centuries, alchemists in different cultural environments attempted to understand and explain chemical transformations by turning to earlier sources of authority: those "philosophers" whose knowledge of alchemical secrets allowed them interrogate the composition of substances, the possibility of material transformation, and the limits and efficacy of alchemical products. As knowledge of alchemy permeated the Eastern Roman Empire, the Islamic lands, and Latin Europe, its theoretical and technical content inevitably changed in line with new contexts. Alchemical theory adapted to accommodate new positions and practices, even as core tenets and the names of earlier authorities were retained. Sometimes they reflected practical observations; at other times, they reflect philosophical concepts or medical terminology.

In this chapter, we trace how alchemical thinking emerged with regard to four major topics, as the tradition itself moved across linguistic and cultural boundaries. First, how did alchemists envisage the physical composition of material substances – the very building blocks of their science? Second, how did theoretical and practical factors contribute to their attempts to classify substances, including metals and other minerals? A grasp of these topics provides the necessary foundation for our third topic: theories of metallic transmutation as they developed from late antique ideas about "dyeing" substances to the fully-fledged elixir theories of Arabic and Latin alchemy. Finally, we turn to an alchemical theory that emerged in the later, Christian Middle Ages: the notion of the "quintessence" as both a perfected form of matter and a sovereign medicinal remedy.

COMPOSITION OF SUBSTANCES

Byzantine and Syriac sources

Late antique and Byzantine natural philosophers, alchemists, and physicians all inherited the ancient Greek theory posited by Aristotle that the primary constituents of the natural world are four elements: earth, air, fire, and water. According to this theory, each element is marked by two qualities, respectively taken from two pairs of opposites: (a) hot versus cold and (b) dry versus moist (see below). As the heirs of Galenic medicine, Byzantine doctors – such as Aëtius of Amida in the sixth century or Paul of Aegina in the seventh century – adopted the same primary qualities to justify and classify the powers of simple drugs in their medical encyclopedias.[2]

The four elements could also be used to explain the composition of minerals when combined with another Aristotelian theory – namely, the idea that there are two exhalations trapped underground: the smoky–dry exhalations that solidify into minerals (e.g. sulfur, orpiment, cinnabar) and the vaporous–moist exhalations that condense into metals (*Meteorology*, III 378a–b; Wilson 2013: 271–7). This theory, often combined with Plato's view of metals as primarily composed of water, was inherited by Neoplatonic and Byzantine scholars, from Olympiodorus in the sixth century to Blemmydes in the thirteenth century, who commented on the *Meteorology* or wrote summaries of Aristotle's natural philosophy.[3]

Traces of the two Aristotelian exhalations surface in Byzantine alchemical writings as well. In his commentary on Zosimos' work, the alchemist Olympiodorus – perhaps to be identified with the homonymous Neoplatonic thinker – provides a doxographical account of the material principles (*archai*) of reality, namely the permanent substances that underlie nature. Many ancient and modern alchemists are said to have recognized a double principle, "vapor" (*atmos*) and "smoke" (*kapnos*), which are intermediaries between the four elements and their qualities (Berthelot and Ruelle 1887–1888: II:83–4; Viano 1995). These natural substances correspond to specific products of alchemical operations: smokes to the "sublimates" (*aithalai*) produced by subliming certain minerals (i.e. those that pass directly from the solid to the vaporous state) and vapors to "moist sublimates" (*aithalai hugrai*) produced through distillatory devices (Berthelot and Ruelle 1887–1888: II:84–5).[4]

In his nine lectures on the making of gold, the seventh-century Byzantine alchemist Stephanus of Alexandria explicitly relates vaporous exhalations to metals. Like Olympiodorus, he writes: "two are the material principles and causes of all things: the vapor that rises up and, among dry (elements), a smoky exhalation … : the vapor is the matter of air, the smoke the matter of fire" (*Lecture* VI 17–19 and 45–6 in Papathanassiou 2017: 187). Stephanus goes further, equating the elemental air, hot and moist, with mercury (*Lecture* V 28–9 in Papathanassiou 2017: 181–2). He describes mercury as a substance mixed within metallic bodies and conceptualizes it as their perfect nature hidden inside (*Lecture* IV 180–7 in Papathanassiou 2017: 179). Indeed, earlier alchemists already insisted on the fact that the liquid metal could be extracted from a variety of mineral substances (e.g. cinnabar, arsenic ores, stibnite), as emerges from a book on the working of mercury, which is preserved in Syriac alchemical manuscripts and ascribed to the third/fourth-century alchemist Zosimos of Panopolis (Berthelot et al. 1893: II: 242–4). The book also includes a short quotation from the ancient Egyptian alchemist Pebichius, who was said to have claimed that "all the bodies are mercury" (Martelli 2013: 245).

Volatile sublimates – and mercury in particular – were therefore seen as important constituents of metals and minerals, usually identified as the substance's "soul," capable of bearing a dyeing *pneuma* (or "spirit"). Metals and minerals, in fact, were often likened to living beings, conceptualized as a mixture of body and soul.[5] A Byzantine alchemist of the late eighth to ninth centuries, usually referred to as the *Anepigraphos*, or Anonymous, philosopher, speaks of metallic bodies as a mixture of *molubdochalkos* (literally, "lead–copper alloy") and mercury; if mercury, when heated, flies away, *molubdochalkos* is described as the matter that remains (Berthelot and Ruelle 1887–1888: II:429–31).

Within a practical context, other metals and substances could also be associated with the prime matter that had to be "dyed" and transformed into silver or gold. The term *tetrasōmia* (literally, "the four bodies"), already defined by Zosimos as the "matter of (metallic) bodies" (Berthelot and Ruelle 1887–1888: II:223–5), was inherited by later alchemists, such as Synesius in the fourth century, and later by Olympiodorus and Stephanus. On the other hand, Egyptians were told to believe that all the base metals were made from lead, at least according to Zosimos' report (Berthelot and Ruelle 1887–1888: II:167–8). This theory was further developed by the alchemist Olympiodorus, who identified black lead as a supporting matter able to receive any dye: because of its blackness, lead already encompasses all the colors and could be further qualified through the application of specific dyeing agents (Berthelot and Ruelle 1887–1888: II:91–6;

Viano 1996: 199; Dufault 2015: 236–7). Black lead is in turn likened to the residues or ashes produced by exposing minerals or metals to fire. These ashes were conceptualized as dead bodies – that is, bodies deprived of their souls or, in other words, of their volatile components – that could be brought back to life through alchemical procedures of dyeing (Papathanassiou 2005: 120–2).

Arabic alchemy

The alchemy that developed from the eighth century onwards in the Arabo-Muslim world is fundamentally indebted to Greek alchemy and borrowed its concepts, especially in relation to theory. Muslim writers therefore adopted the Greek theory of the elements and elementary properties, sometimes in its Galenic form, which classifies properties according to four degrees of intensity, most often in a slightly simplified form. Usually, things were classified according to the degree of intensity of the two most intense properties, one for each of the two pairs: hot–cold and dry–humid (Harig 1974).

This Greek theory gave rise to the conception of "occult" and "manifest" properties. Substances are characterized by two groups of opposite properties. The two most intense properties (e.g. cold and moist) are those that are perceived by the senses, and therefore manifest. The other two (in this example, hot and dry) are therefore occult, or hidden from the senses. This led some alchemists to go further and to consider the manifest properties as *external* to a substance and the occult properties as *internal*. This allowed them to group metals two by two, according to pairs of properties: gold, which is hot and dry, is thus the opposite of lead, which is cold and wet. It follows that gold has lead inside itself, and vice versa. According to this approach, the aim of the alchemists is therefore to change metals by making their occult properties manifest and the manifest properties occult by bringing out the internal qualities – thereby changing lead into gold. Occult properties were sometimes called the spirit or soul of the metal and manifest properties the body.[6]

The theory of sulfur and mercury developed in the Islamic East as an extension of the Aristotelian theory of the double exhalation and its continuations (see above and Norris 2006) that reflected the important place of mercury and sulfur in speculations about the nature of matter. This conception identified Aristotle's watery exhalation with mercury and his earthy exhalation with sulfur – mercury and sulfur thus became the two principles of metals. The theory appeared very early: it is already found in the ninth-century *Book of the Secret of Creation* (*Kitāb Sirr al-Khalīqa*) by Pseudo-Apollonius of Tyana, in an ancient form in which mercury contains its own sulfur, which it reacts with, thereby coagulating into the various metals.

The most common form in which this theory appeared among Arabic authors is that described in *fann* 5 of the *Physics* (*Ṭabīʿiyyāt*) of the *Book of Healing* (*Kitāb al-Shifāʾ*) by the great Persian philosopher Ibn Sīnā (Avicenna).

The metals are composed by the coagulation of the two exhalations, identified as mercury and sulfur, and the metals differ from one another according to the proportion of sulfur and mercury, their degree of purity, and the time of their cooking in the bowels of the earth, which is counted in hundreds of years. It was also in this form that the theory penetrated the Latin West at the time of the Arabic–Latin translation movement during the twelfth and thirteenth centuries.

It should be noted here that the theory of elements and elementary properties inherited from the Greeks has, for some authors, been developed and made more complex. A major example of this can be seen in the theory of the composition of substances in the texts of the Jābirian corpus. These texts develop a method for calculating the proportions of elementary properties of a complexity rarely equaled: the balance of the letters (*mīzān al-ḥurūf*), also called the science of the balance (*ʻilm al-mīzān*). In this system, the elementary properties are calculated directly from the names of things; the letters of the Arabic words for things correspond to amounts[7] of elementary properties and thereby allow one to know their composition. Describing the method of calculation in detail would take up too much space here, but the refinement of this doctrine deserves special attention.[8] The balance of the letters opened the door to the elixir theory, as discussed below. The influence of Jābirian texts on Arabic alchemical literature also allowed a wide diffusion of the science of the balance. On the other hand, it never penetrated the West because of its close link with the Arabic language, which was impossible to transfer into Latin.

Alchemy in Latin Europe

The translation of Arabic alchemical sources into Latin began, if we are to believe the information given in the colophon, in 1144 when the *Book on the Composition of Alchemy* (*Liber de compositione alchimiae*) was written. The translator, possibly Robert of Chester, knew that Latin readers would be unfamiliar with the novel subject matter expressed in this treatise, a translation of the *Epistle of the Wise Monk Maryānus to the Prince Khālid b. Yazīd* (*Risālat Maryānus al-rāhib al-ḥakīm li-al-amīr Khālid b. Yazīd*), a dialogue between a Greek monk and an Islamic prince (Dapsens 2016; Moureau 2020: 116) that provides a fitting metaphor for the transmission of alchemical knowledge between cultures.

The choice of Arabic texts to be translated was apparently not particularly systematic, resulting in some shifts in theoretical emphasis. For instance, since only a few items from the Jābirian corpus, such as the *Book of the Seventy* (*Liber de septuaginta*), were rendered into Latin, Jābirian doctrines tended to be communicated through other texts, such as the influential *De anima* discussed below. While adherents of alchemy at first relied heavily on these translations, over the course of the thirteenth century they began to adapt their source texts, making systematic attempts to raise the prestige of alchemy by showing that it rested on generalizable, natural principles, including those of Aristotelian natural philosophy and Galenic medicine. As practitioners, they also used their deep knowledge of the properties of chemical substances to build on existing theories of material change, contributing to a philosophy of metallic generation and transformation that would become the established position in medieval Europe – even as alchemy itself failed to gain a foothold in the universities.

Some of the main positions gleaned from Arabic sources were surveyed in the *Liber mineralium* of the thirteenth-century Dominican philosopher Albert the Great. Albert adjudicates between different theories about the composition of metals, although his arguments are not based on practical experiments. For instance, Avicenna expressed skepticism toward the idea of transmutation in his *Book of Healing*, arguing that unless the differences of a thing in *species* (such as lead) were known, one could hardly transmute it into something else (such as gold). Albert's own response was cautiously positive: rather than transmuting entire *species*, the alchemists sought merely to change the specific form of a given metal – a task that he considered achievable, albeit only with considerable difficulty, since it would still require a new form to be imposed on matter (Newman 1989: 431–2). He also rejected several theories on the composition of substances promoted by alchemists, especially one named "Callisthenes" – probably an erroneous reading of Khālid b. Yazīd (Wyckoff 1967: 172). According to this view, the specific form of all metals is gold, and other metals are therefore merely imperfect, or "diseased," iterations of gold. By implication, base metals can be encouraged to ripen or digest into gold using alchemical arts. A similar view had been stated centuries earlier in the *Book of the Secret of Creation* of Pseudo-Apollonius, and in fact would continue to inform practitioners' arguments well into the sixteenth century.

When it came to theorizing on the composition of substances for the purposes of alchemical practice, one of the most influential treatises of the Latin Middle Ages was the *Summa perfectionis magisterii* (*Compendium of the Achievement of the Magistery*) pseudonymously attributed to Jābir b. Ḥayyān, probably composed toward the end of the thirteenth century (Newman 1991). The Latin "Geber" is interested primarily in gold-making, describing three "orders" of medicine that produce successively more convincing iterations of gold – from a superficial change of color to actual transmutation. While Pseudo-Geber builds on the bedrock of the sulfur–mercury theory, he approaches the theory in a novel way by conceiving of the two principles in terms of small, homogeneous, and closely packed particles, or *minima*: a solution that can be taken as an early example of a corpuscular theory of matter (Newman 1991; Newman 2006). He also considered the metals to be made up primarily of a very pure version of mercury (here in the sense of elemental mercury) that had been altered by sulfur and "earthy" matter. Since mercury contained its own, internal sulfur, it did not require further admixture – thereby establishing it as the fundamental prime matter of the alchemical work. This theory, which historians have designated "mercury alone" or "mercurialist," would become a staple of Latin treatises throughout the fourteenth century and beyond – although, as we shall see, the term "mercury" was itself open to interpretation.[9]

THE CLASSIFICATION OF SUBSTANCES

In the early Byzantine period, alchemists credited the first-century Pseudo-Democritus, one of the fathers of alchemy, with a classification of substances according to the four elements (Berthelot and Ruelle 1887–1888: II:78). This criterion is no longer observable in what remains of Pseudo-Democritus' catalogs, which open his books on the making of gold and silver,[10] but it does match Byzantine practices of classification. For instance, in his fifth lecture on the making of gold, Stephanus of Alexandria associates each element with both a bodily humor and a specific substance: thus air is similar to blood and mercury, fire to yellow bile and copper, earth to black bile and scoriae, and water to phlegm and gold. He then goes on by dividing the twelve signs of the zodiac into four groups, each one linked to a primary element (*Lecture* V 35–104 in Papathanassiou 2017: 182–4). Stephanus claims that both planets and metallic bodies (as well as their colors) depend on the zodiac (Papathanassiou 2006: 175–80), thus elaborating on the association between metals and planets that emerges from different lists handed down in Byzantine and Syriac alchemical manuscripts. For instance, the tenth-century MS *Marcianus* gr. 299 includes a list that perfectly matches the classification attributed to Proclus in the *Commentary on Aristotle's Meteorology* by the Neoplatonic philosopher Olympiodorus: (a) sun–gold, (b) moon–silver, (c) Saturn–lead, (d) Jupiter–electrum, (e) Mars–iron, (f) Venus–copper, and (g) Mercury–tin.[11]

Byzantine authors also made frequent use of the alchemical egg as a classificatory device.[12] The practice of distilling eggs, already described in Zosimos' writings, became popular in later alchemical texts (Mertens 1995: 202–3). On the one hand, the substances produced by distilling eggshells, egg whites, and yolks were assimilated to the four elements, as clearly emerges from the anonymous text referred to as *The Work of the Four Elements*. Here, fractional distillation is used to extract the four elements from eggs, in a sequence that has telling similarities with some Jābirian texts (Berthelot and Ruelle 1887–1888: II:337–42; Colinet 2000a: 165–90). On the other hand, each part of the egg was linked to substances that cannot be identified with distillation products. An anonymous text entitled *Names of the Egg* deals with "the egg made of the four elements" (*ōon tetrastoichon*) and lists the following associations: (a) eggshell–earth–metals, namely copper, iron, tin, and lead; (b) egg white–divine water, along with other alchemical waters; (c) egg yolk–presumably air (even though not mentioned in the text)–malachite along with various minerals, such as ocher and soda; and (d) the oily part of the egg–fire (Berthelot and Ruelle 1887–1888: II:20–1).[13] Likewise, in his seventh/eighth-century treatises on the making of gold, the so-called philosopher Christianus divided both natural substances and alchemical operations into four main classes, which corresponded to the four parts of egg: shell, membrane, white, and yolk (Berthelot and Ruelle 1887–1888: II:409–10).[14]

An operational basis is also detectable in a different classification of substances established by Zosimos. This classification emerges in early Byzantine commentators on his works as well as in Syriac books ascribed to him. The first of these Syriac books clearly divides minerals into three classes, according to their resistance to high temperatures: (a) *sulfurs* – namely sulfur, orpiment, and realgar – which quickly flee (i.e. evaporate) when heated; (b) *sulfureous substances*, which slowly evaporate (identified with "all the valuable stones that come out of earth, from mercury to malachite"); and (c) *bodies* – namely, copper, iron, tin, and lead – which do not evaporate when exposed to fire (Berthelot et al. 1893: II:214). The same categories are used by the alchemist Olympiodorus, who further commented on this threefold classification (Berthelot and Ruelle 1887–1888: II:75; Martelli 2014a: 206–7).

Zosimos' Greek writings account for a simplified division, which only distinguishes between *bodies* (*sōmata*), corresponding to the third class mentioned above, and *incorporeal substances* (*asōmata*), which seem to include substances belonging to the first and second class (Berthelot and Ruelle 1887–1888: II:196; Martelli 2014b: 32–3). When exposed to fire, these substances evaporate and produce different kinds of *sublimates* (*aithalai*). The opposition of bodies versus nonbodies, as well as sulfurs (*ta theia*), sulfureous substances (*ta theiōdē*), and sublimates, was inherited by later alchemists, such as Stephanus of Alexandria, the authors of four iambic poems on alchemy, and the so-called *Anepigraphos* philosopher. Similar classifications also left their mark in the *ouroboros*, the snake eating its own tail often depicted in alchemical manuscripts: its four legs were interpreted as a visual representation of the four metallic bodies, while its three ears represented three kinds of sublimates (Berthelot and Ruelle 1887–1888: II:22–3).

Various Greek classifications were transmitted to the Arabo-Muslim world, notably through translations from Greek into Arabic. However, a classification of substances quickly became established in Arabic alchemy: the distinction between bodies and spirits, sometimes with the addition of souls. This theory spread but underwent many variations depending on the authors; it remained stable only in its principles. In the *Book of the Secret of Creation*, this classification had not yet become established, but the author already made a distinction between the body of materials and their spirit (see above). In the Jābirian corpus, the doctrine often varied from one text to another.

Different classifications were proposed, but all of them were variations of the body–spirit classification. Thus, in the *Great Book of Properties* (*Kitāb al-khawāṣṣ al-kabīr*), three categories were given: substances that sublime are referred to as spirits, and all other materials are referred to as bodies, which divide themselves into fusible bodies (i.e. metals) and other bodies (Kraus 1942: 18–20). For the author of the *Book of Mercy* (*Kitāb al-raḥma*), substances can be either bodies or spirits, but they are all composed of both: they are called body or spirit according to what predominates in them, but every body has spirit in it and every spirit has body in it (Berthelot et al. 1893: 3:146). In the *First Book of the Element of the Foundation* (*Kitāb usṭuqus al-uss al-awwal*), in addition to bodies as they are usually described (fusible and nonfusible), spirits are divided into two categories: spirits, which are not oily and include mercury, sal ammoniac, and camphor; and souls, which are oily and include sulfur, oil, and arsenic (Holmyard 1928: 67–8). It is not necessary to multiply examples, since these principles – body and spirit – are the only ones transmitted from text to text.

The Arabo-Muslim distinction between spirits and souls is subtly different from that found in Greek texts. Spirits are substances that sublime; that is to say, they can pass into a volatile state with or without fire. Bodies are not volatile, and are further distinguished into substances that are fusible (namely metals) and nonfusible. The meaning of the term "soul" varied, however – too much to be generalized here. These classifications were superimposed on the assumption that the things of the mineral kingdom had a body, the fixed part, which was manifest and passive, and a spirit, the volatile part, which was occult and active. The category of nonfusible bodies was also often divided into subclasses, including salts and stones. One of the richest classifications of substances that has come down to us is that of the great ninth/tenth-century physician Abū Bakr al-Rāzī (ca. 250–313 or 323/854–925 or 935). In his *Instructive Introduction* (*Madkhal ta'līmī*) and *Book of Secrets* (*Kitāb al-asrār*), Rāzī elaborated a refined classification (Stapleton et al. 1927). Although not especially innovative, this classification was extremely detailed – distinguishing between four spirits, seven bodies (metals), thirteen stones, six vitriols, six boraces, and eleven salts.

The body–spirit classification was transmitted to the Latin world through Arabic–Latin translations. Thus, the main works on which the Latin alchemists based their work presented a body–spirit classification, which consequently became widespread in the first centuries of Latin alchemy. The classification of Rāzī was also translated, and it influenced several Latin authors (Newman 1991: 109–42). For instance, *The Right Path* (*Semita recta*) of Pseudo-Albertus Magnus distinguishes between the "four spirits" that color metals (mercury, sulfur, auripigmentum or arsenic, and sal ammoniac) and those that do not add color but nonetheless serve as instruments to the work in other respects (such as salts, alums, vinegar, and urine) – exactly the kind of substance that would later be spurned as proper ingredients by Latin alchemists in the mercurialist tradition (Rampling 2020: 38–42). As elixir theory developed, the classification of animal, vegetable, and mineral ingredients was also more finely honed, as alchemists pondered a new, fundamental distinction – between substances safe for use in medicinal elixirs and those that were toxic to the human body.

DYEING, TRANSMUTATION, AND ELIXIR THEORY

Theories of dyeing: Greek sources

In his *Letter on the Making of Gold* addressed to the patriarch of Constantinople Michael Keroularios, the eleventh-century Byzantine philosopher Psellos defines alchemy as "the art of fire" (*empurios technē*), which acts on the elemental composition of various materials, such as metals and stones. Alchemical dyes are set in the wider framework of natural philosophy and explained by considering the transformations of the four elements, such as evaporation and condensation of water in meteorological phenomena.[15] A similar situation emerges in a recipe book ascribed to the thirteenth-century Byzantine philosopher Blemmydes, who describes a complex technique for distilling eggs. According to his introduction, the process was meant to manipulate the four elements, among which earth was likened to eggshell (Berthelot and Ruelle 1887–1888: II:452). Indeed, the four elements, Blemmydes explains, constitute the "stone of the sages" (*lithos tōn sophōn*), namely the ultimate transmuting agent that the author wanted to produce. This dyeing substance is described as a deeply purple, dry powder – *xērion* in Greek – that, when added to melted silver, can transform it into gold (Berthelot and Ruelle 1887–1888: II:457). The term, transcribed in Syriac letters (*ksīrīn*), appears in many Syriac alchemical recipes with reference to dry dyeing substances, which were often produced by treating and distilling eggs (Berthelot et al. 1893: II:42–55).

Dry substances were certainly used as dyeing agents since the earliest phases of Greek alchemy. However, the concepts of the philosophers' stone and *xērion* (or elixir) seem to have been fully developed only in the late Byzantine tradition (Berthelot and Ruelle 1887–1888: II:446–50). Byzantine alchemists, such as Stephanus and the *Anepigraphos* philosopher, inherited a distinction that the late antique alchemist Synesius traced back to Pseudo-Democritus himself: Egyptians used to roast metals, while Persians preferred using liquid dyes (Martelli 2013: 238–9). Among dyeing liquids, the so-called "divine water" or "water of sulfur" represented the alchemical water *par excellence*, whose role was at the core of Zosimos' alchemy and continued to be highly praised by all Byzantine alchemists. They speculated on its nature, on the unity or plurality of its species, and on its capacity for producing stable dyes[16] – often conceptualized as "deep tinctures" (*katabaphai*) – in contrast to superficial and unstable dyes.

Divine water was often likened to mercury, which was credited with a key role in dyeing procedures. The philosopher Christianus, for instance, quotes Zosimos to confirm the identification of divine water as mercury (Berthelot and Ruelle 1887–1888: II:397). As we have already seen, the Syriac books ascribed to Zosimos classify substances according to their resistance to fire. Those ingredients that easily evaporate (i.e. sulfurs and sulfureous substances) produce unstable dyes; however, if mixed with mercury, both the volatile substances and the dyes they produce become more stable (Berthelot et al. 1893: II:214, 242–4; see Martelli 2014b: 35–9). This mechanism, already described by Synesius, is explained again by Olympiodorus (Berthelot and Ruelle 1887–1888: II:90).[17] Mercury acts as a medium: because of its liquid nature, mercury absorbs and holds the souls (i.e. the volatile parts) of minerals that evaporate when heated; thus, it incorporates their dyeing properties and becomes their support. Finally, when mixed with the metallic matter to be dyed, mercury can transfer these dyeing properties and bring about a stable tincture.

More generally, the divine water was identified with the key transmuting agent that, if made white, could transform metals into silver and, if made yellow, into gold. This principle followed a quite standard chromatic sequence of the different steps of the alchemical transformation. Stephanus, for instance, quotes an earlier alchemical saying: "After the cleaning of the copper and its attenuation and blackening before the latter whitening, then is the stable yellowing" (*Lecture* II 67–9 in Papathanassiou 2017: 163). If blackness was often identified with the color of the phase in which metals or minerals were reduced to ashes, these ashes were then treated by liquid substances: these substances, as carriers of a dyeing and revivifying pneuma (see above), could progressively make the ashes white and yellow – that is, transform them into silver and gold.

Elixirs in Arabic alchemy

The concepts of dyeing and transmutation passed from the Greek world into Arabo-Muslim culture. However, the possibility of modifying the properties of a metal so as to transform it into another metal was not accepted without question. Indeed, it was often at the center of controversy. Great philosophers such as Avicenna opposed it; as we have seen, he considered transmutation to be highly unlikely (Moureau 2016a: 11–18). One of the major concerns of the Arabic alchemists was how to change a metal in all its properties, not just some of them, and how to make the new properties durable. Here the idea was not just to dye the metal, although the term dyeing (*ṣibgh*) was still often used, but to transform the entire nature of the metal. Temporary changes were not therefore considered to be transmutations.

In the first centuries of Arabic alchemy, the Greek concepts described above circulated, and it was not until the writing of the corpus of texts attributed to

Jābir b. Ḥayyān that one doctrine tended to take precedence over the others: namely, the theory of elixirs (Kraus 1942: 1–18). Borrowed from the Greek tradition (the word itself comes from Greek through Syriac; see above), these "medicines" for metals were developed and popularized in the Arabo-Muslim world. They were dry powders that could transmute a base metal into gold or silver. This theory, closely related to the science of the balance, was based on the idea that all things were composed of the four elements and differed according to their proportion of elementary properties (hot, cold, dry, wet). According to these authors, it is possible to change these proportions in things, and thus to give lead the proportion of gold, thereby transforming it.

It is here that the elixirs came into play, by modifying the elementary proportion of things in order to transmute them. How were these elixirs produced? The alchemist took a "stone" – that is, a material (not necessarily a real stone) – from which the elixir was made (Vloten 1968: 265–6; English transl. in Stapleton et al. 1927: 367). The identification of this "stone" would be the subject of many pages in Arabic alchemical literature. It was sometimes said that its proportion of elementary properties had to be as balanced as possible. The alchemist then had to distill this material in order to isolate its four elements. Next, each element was prepared in such a way that it would be characterized by only one of its two elementary properties: thus, fire would become pure heat, water pure coldness, air pure humidity, and earth pure dryness. The practitioner could then mix the four, isolated properties in a very precise proportion, such that their product would be able to modify that of the base metal – this crucial proportion being calculated from those of the base and the desired metal. This mixture was called elixir. It was projected (or cast) onto the prepared base metal and, if all went well, transmuted it into gold or silver.

Elixir theory in Latin Europe

The theory of elixirs had an important influence in the Arabo-Muslim world, and also penetrated the Latin West in various forms. Its strictly Jābirian form spread through texts such as the Latin translation of the Jābirian *Book of the Seventy*, among others, and the pseudo-Avicennan alchemical *On the Soul* (*De anima*). For instance, *De anima*, a collation of three initially separate works written in al-Andalus that were translated into Latin during the thirteenth century, was one of the major sources on alchemy used by Vincent of Beauvais in his encyclopedic *Speculum naturale* (Moureau 2012). A different version of Jābir's theory was conveyed by another Arabic work, the pseudo-Aristotelian *Secret of Secrets* (*Secretum secretorum*), which presented the elixir as a compound of elementary properties in equal proportions, implying that the elixir itself had a universal application, acting more like a "ferment" (see below), transmitting its own qualities to the metal (Moureau 2013).

Both the *De anima* and the *Secretum secretorum* were studied closely by the English Franciscan Roger Bacon, and played an important role in shaping his own elixir theory (Newman 1994a; Newman 1995; Moureau 2013). Bacon regarded alchemy as a fundamental science of nature, since its subject was the elements from which all other substances derived, organic as well as mineral.[18] This approach was, accordingly, rather more ambitious (and contentious) than that of his near contemporaries Albert the Great and Pseudo-Geber, both of whom treated alchemy as the part of natural philosophy primarily concerned with the transmutation of metals. Bacon, on the other hand, viewed alchemy as an important tool for prolonging life. Rather than focusing exclusively on mineral substances, he followed *De anima* in identifying the stone with organic products, especially blood.

This is not to say that Bacon received his Arabic–Latin sources uncritically. For instance, *De anima* sets out three prime materials: the animal stone (blood), the vegetable stone (hair), and the natural stone (eggs). Through a process of fractional distillation, each of these "stones" is reduced to its four constituent elements, which are then prepared according to the Jābirian theory described above, so as to isolate the properties of hot, cold, wet, and dry before recombining them in proportions that are carefully calibrated to "rebalance" the proportions already present in a given base metal. Once combined with a "ferment" made from a precious metal, these elixirs can be used to transmute the target metal into gold or silver (depending on the number of elements used and the nature of the ferment; Moureau 2013: 286–92). In Bacon's interpretation, the best starting matter was blood, but, rather than separating it into four elements, he proposed a system based on the four Galenic humors. Blood is one of the four *compound* humors; from this compound blood, Bacon proposed separating out four *simple* (or more essential) humors, which then took the place of the pseudo-Avicennan elements in the preparation of the elixir.

Thus far, Bacon follows *De anima* in the notion that an elixir must furnish exactly the opposite proportion of elements to that of whichever base metal is to be transmuted. However, Bacon also envisaged using elixirs to prolong life – and here he did not propose calculating the proportion of elements to suit individual human complexions. Rather, he conceived of a single elixir in which all the elements are equally proportioned, which can then impose its own, perfect equality on the human body, regardless of the individual's complexion – a system closer to the ferment idea than to either Jābirian elixir theory or traditional Galenic medicine (Moureau 2013: 316–19).

This conception of the elixir as an equally proportioned body with universal application seems to have resonated with later writers, since the idea of an equally balanced, universal elixir recurs in some of the most influential treatises of the fourteenth century.[19] The result, in fact, would be an amalgam of two traditions: pseudo-Geberian alchemy, with its focus on metallic composition and transmutation, and an approach that Michela Pereira has termed "alchemy of the elixir," which proposes a multipurpose elixir that both transmutes metals and heals human bodies (Pereira 1995b).

Within this amalgamated tradition, Pseudo-Geber's views dominated with regard to transmutation. The *Summa perfectionis* actually condemns the use of organic ingredients (such as blood) and nonmetallic minerals (such as vitriols and salts) that, although sometimes useful for preparing metals, do not themselves constitute the stone, which is instead made from mercury. Pseudo-Geber's mercurialist theory (together with the critique of nonmetallic ingredients, which soon became standard) was blended with the idea of a universal transmuting and medicinal elixir in a series of influential treatises pseudonymously attributed to the Montpellier physician Arnau de Villeneuve (or Arnald of Villanova), the Majorcan philosopher Ramon Llull (Raymond Lull), and the English monk John Dastin.[20]

For instance, the foundational work of the pseudo-Llullian corpus, the *Testamentum*, written in 1332, delineates the scope of the elixir in a passage very similar to definitions given in two other influential works, both entitled the *Rosary of Philosophers* (*Rosarius philosophorum*) – one being the core work in the pseudo-Arnaldian corpus, the other usually being attributed to Dastin. These passages emphasize the value of an equally proportioned elixir not just for transmuting metals, but also for healing all the illnesses that assail the human body, for, according to the *Testamentum*, "it is of the subtlest and noblest nature, and it reduces everything to the utmost balance ... therefore don't wonder if this remedy was sought for more eagerly than any other, because in it all other remedies are encompassed" (Pereira 1998: 30). While mineral ingredients are privileged in the theoretical part of the treatise, the subsequent practical books do refer to some organic ingredients, including what seems to be an early example of the alchemical use of distilled wine.

In general, however, the philosophical and practical content of these "elixir" texts is far more geared toward mineral preparations, specifically involving three metals: mercury, gold, and silver. In his own *Rosarius*, for instance, Pseudo-Arnald states that ferments are necessary to transform the stone into elixir by imparting the form of the precious metals to the stone. Much medieval Latin alchemy was predicated on exactly this two-step approach: the preparation of the stone from quicksilver or a base metal, followed by its fermentation with gold (for the red work) or silver (for the white) to make the elixir – the artificial agent of transmutation. In a common analogy with bread-making, the stone takes on the role of dough, with the gold and silver acting like yeast. The resulting elixir is so purified and digested that it exceeds the perfection of normal gold and silver; indeed, it is "more than perfect" (*plusquam perfectum*). When projected onto an imperfect metal like lead, this super-perfect elixir will raise it to the "normal" perfection of precious metal (Rampling 2013: 61–3).

The pseudo-Llullian *Testamentum* offers a similar approach, ordaining that "the ferment is not extracted unless from Sol and Luna [i.e. gold and silver], since we require nothing except that the stone may be converted into its like" (Pereira and Spaggiari 1999: 172). Another influential pseudo-Llullian treatise, the *Book of the Secrets of Nature, or Quintessence* (*Liber de secretis naturae, seu quinta essentia*), relates how fermentation with precious metals fixes the fluid stone into a solid; a process that simultaneously confers the form of gold and silver upon the matter of the stone, creating an elixir. In this approach, mercury provides the "material" principle of the elixir and gold and silver the "formal" principle.

Not all Latin authorities, however, agreed that precious metals were necessary to ferment the work. For instance, the adept Guido de Montanor seems to have taken the view that other substances besides gold and silver might serve as ferments, provided they were fixed – a view later explored by the English canon George Ripley, who was concerned about the cost of transmutation for practitioners who could not afford gold (Rampling 2020: 111–18). Despite their seeming prohibition in mercurialist rhetoric, organic ingredients such as eggshells, blood, and – most of all – distilled wine continued to feature in alchemical theory and practice throughout the fifteenth century. Indeed, the use of diverse ingredients actually gained in legitimacy as the pseudo-Llullian corpus expanded to incorporate a new concept: that of the "quintessence."

THE QUINTESSENCE AND ALCHEMICAL MEDICINE

The term *quinta essentia*, or fifth essence, originally denoted Aristotle's fifth element: the ether that composed the entire celestial region. The concept is present in a few rare Arabic alchemical texts, but did not play as important a role as it did in the Latin world. It appears notably in the Jābirian corpus, but in a cosmological context, where it is called the fifth nature (*al-ṭabīʿa al-khāmisa*). Here it does not mean the same thing as the Aristotelian quintessence: it is more a fifth property than a fifth element, and no alchemical use was made of it (Kraus 1942: 153).

It was in fourteenth-century Europe that the idea of the fifth element emerged as a major component of alchemical theory, thanks to the confluence of two streams: the quintessential medicine of John of Rupescissa and pseudo-Llullian alchemy. The "fifth element" first appears in the keystone of the pseudo-Llullian corpus, the *Testamentum*, as the pure, primordial substance from which the four terrestrial elements are created. These four elements retained the purity of their parent element until the Fall of Man, when nature was corrupted by sin. Since that time, the quintessential remnants have lain buried within the grosser stuff of the four corrupted elements – to be liberated either on Judgement Day or through the art of the alchemist (Pereira 1992: 184–7; Pereira and Spaggiari 1999: 12, 14). However, while the theory provides a parable for the salvific power of alchemy (and its role in extracting a pure, active virtue out of matter), it is not clear that the writer intended it to be taken as a literal history of material composition, rather advising that his words should be interpreted as "philosophical speech" (*sermo philosophalis*; Pereira and Spaggiari 1999: 14).

The quintessence first appeared as a tangible alchemical product in the *Liber de consideratione quintae essentiae* of the Franciscan visionary, John of Rupescissa. Writing around 1351/2, Rupescissa described the quintessence as a novel medicinal preparation made by distilling wine, to yield a clear, incorruptible liquid in which the primary qualities of the elements had achieved perfect balance (Halleux 1981; DeVun 2009). By naming his elixir "quintessence," John invoked a relationship with the incorruptible Aristotelian ether, claiming that the elementary qualities of the quintessence were so perfectly proportioned that it could expel corrupt matter from the human body. Given that Aristotelian physics treated the terrestrial and celestial regions as entirely separate, this association could never be more than analogical. Rupescissa contrasted this equally complexioned medicine with conventional medical remedies, the efficacy of which was dictated by the relative intensity of their elemental primary qualities (a process similar to that discussed above in relation to Jābirian alchemy). The quintessence could furthermore be used as a solvent to extract the healing virtues out of animal, vegetable, and mineral substances, giving rise to an entire alchemical pharmacopoeia of quintessences, each a hundred times more medically efficacious than its parent substance.

Rupescissa's quintessence became inseparably linked to pseudo-Llullian alchemy later in the fourteenth century, when parts of his treatise were incorporated into a new, longer alchemical work, the *Liber de secretis naturae, seu quinta essentia*. The author of this work wrote pseudonymously under the name of Llull and created an aura of Llullian authenticity by adding an elaborate preface and conclusion, as well as Llullian terms and diagrams. In both form and content, the book was a work of synthesis: thus, John of Rupescissa's treatise provided the grist of the first two books (or distinctions) on medicinal alchemy, while the *Tertia distinctio* draws on the alchemy of the *Testamentum*, with the important addition of quintessence of wine, now to be used in transmutation as well as healing.

This fusion of two theoretical strands – the more traditional, elixir-based alchemy of the *Testamentum* and Rupescissa's new quintessential medicine – contributed to a new and influential focus on alchemy as a multipurpose art, intended to heal both metals and bodies. Writers in this tradition moved away from the notion of a single, universal elixir in favor of an array of animal, vegetable, and mineral stones (labels already familiar, although in a different context, from Pseudo-Aristotle, *De anima*, and the pseudo-Arnaldian corpus) to differentiate between diverse alchemical products. For instance, the pseudo-Llullian *Letter of Abbreviation* (*Epistola accurtationis*) describes a compound water (*aqua composita*) made by mixing the ingredients for the transmuting mineral stone (including vitriol) with those of the wine-based quintessence, or vegetable stone. This compound water is said to dissolve gold more easily than the vegetable water alone, since gold shares the vitriol's mineral nature, and its parts will therefore cleave more readily to those of the vitriol than to the spirit of wine (Manget 1702, 1:865).

While distilled wine had already been promoted as a remedy in medical *consilia* and pseudo-Arnaldian alchemical writings, Rupescissa was the first to claim it as the basis for an alchemical elixir, explicitly contrasting its medicinal virtues with the toxic nature of "alchemical" (i.e. transmuted) gold. This conception contributed to rising interest in the pharmaceutical value of extracting virtues or essences from animal, vegetable, and mineral materia medica; an influence apparent in the profusion of distilled waters encountered in fifteenth-century recipe books and eventually disseminated in the printed manual of Hieronymus Brunschwig (Brunschwig 1500). By the end of the fifteenth century, the quintessence and its cover names (notably Rupescissa's term *coelum*, "heaven") were ubiquitous in Latin and vernacular alchemical texts.

However, as the properties of the "spirit" of wine lost their initial mystique, readings of the pseudo-Llullian quintessence also changed, broadening to include both metallic essences and a range of "vegetable" solvents. For instance, the attention to metallic extractions within pseudo-Llullian texts gave rise to the idea – widespread in fifteenth-century alchemical commentaries – that essential "mercuries" (or "souls") could be extracted from metals. These "mercuries" (referred to in the *Liber de secretis naturae* as resolutive menstrua) did not equate to the mercurial principle in sulfur–mercury theory, nor to elemental quicksilver. Rather, they signified the active and essential part of the metal that, once separated from its grosser matter, might serve as a solvent for further operations. For instance, a "mercury" drawn out of lead, when combined with a vegetable solvent (often called quintessence, although likely to refer to other solvents in practice), could be used to extract the more precious mercuries of gold and silver (Rampling 2020).

Throughout the fifteenth century, alchemical theorizing continued to be both syncretic and pluralist. During the 1470s, George Ripley popularized pseudo-Llullian theories in Latin and English. His *Marrow of Alchemy* (*Medulla alchimiae*) adopts the *Epistola*'s notion of three distinct stones: a mineral stone heavily influenced by Pseudo-Geber; a vegetable stone that draws on Pseudo-Llull and Guido de Montanor; and an animal stone that hints at the ongoing influence of the pseudo-Avicennan *De anima* (Rampling 2020). In addition, he inserts a discussion of the *Epistola*'s compound water into his chapter on the mineral stone. This eclectic distillation of alchemical sources shows that Ripley, like many other practitioners who paid lip service to the theory of "mercury alone," in fact took mercury to denote a wide spectrum of *materia alchemica* – animal and vegetable, as well as mineral. Thus, in Ripley and his contemporaries, one sees a blending together of strands of alchemical doctrine whose history can be traced from ancient Egypt and Greece, through Syriac and Arabic, then into Latin and eventually the European vernaculars.

CHAPTER TWO

Practice and Experiment: *Alchemical Operations in the Middle Ages*

SÉBASTIEN MOUREAU AND NICOLAS THOMAS

INTRODUCTION

Practices and experiments in alchemy are composed of processes conducted in order to achieve definite aims.[1] The main aim of alchemists was making gold (chrysopoeia) or making silver (argyropoeia). However, if the definition of a chemical process is clear and precise in today's chemistry, this was far from being the case in the Middle Ages. Medieval scholars did not think in terms of chemical reactions as modern chemists do, but in terms of transformation: alchemy in the Middle Ages was a science of the transformations of things, and this implies that a discussion on medieval alchemical processes belongs to both chemistry and physics. Medieval alchemists considered that any change of matter – they would say of elements or elementary properties – is alchemical. For instance, they often quoted the process of boiling water, considering that water changes into vapor, which was thought to be air, and therefore that boiling water is the transformation of water into air.

When describing alchemical processes, it is important to distinguish clearly between craft and alchemy. Indeed, craftsmen and alchemists used many processes in common, such as sublimation, distillation, or refining metals, but they differed considerably in their aims. While craftsmen sought to produce materials or artifacts, the main goal of alchemists was to produce an incorruptible metal from a corruptible one (i.e. to manage the transformations of materials and be able to lead them to their perfection). This also implies a clear-cut difference in their approaches: craftsmen conducted processes without the need of understanding them, while alchemists could not but try to comprehend thoroughly the processes they undertook. The theoretical part of alchemy was essential to alchemists, but the theoretical part of the processes implemented in craftsmanship was unknown to most craftsmen in the Middle Ages. If some processes were specific to craftsmen, such as roasting ores to extract metals or casting and hammering metals in order to make artifacts, there were numerous contact points between various crafts (metallurgy, dyeing, etc.) and alchemy in their techniques and processes, and alchemists sometimes did not hesitate to insert craft recipes and considerations into their texts. But alchemists were not craftsmen as such. The practices and experiments specific to craft will be described in Chapter 6.

Studying alchemical processes faces various difficulties. A process is in fact extremely hard to describe in its entirety, and medieval texts are punctuated with information gaps. Tacit knowledge is required to read alchemical texts, and one can hardly imagine a medieval alchemist being able to reproduce any recipe without having access to this implicit information. Moreover, orality or observation of craftsmen or other alchemists – namely, a direct contact with practitioners – was a significant part of alchemy, as well as practice. Reading a recipe in itself entails interpretation. But, in addition to this, there is a portion of knowledge in a recipe that cannot be textually transmitted: know-how, gestural knowledge, or embodied knowledge, however we call it. Indeed, even knowing everything about a process is sometimes not sufficient to conduct it if it requires practical skills: being the best expert on Bach's first partita for solo violin is not sufficient to play it; embodied knowledge is essential.

Another problem lies in the system of written transmission of knowledge. Indeed, knowledge can be passed on by copyists who did not necessarily possess this knowledge: not all scribes were alchemists. Thus, in addition to the usual changes due to manuscript transmission, there were many changes in recipes due to the misunderstanding of the copyists. A recipe is sometimes unrecognizable after being copied only a few times. On the other hand, copyists who had the knowledge necessary to understand recipes, or who believed they have it, did not hesitate to modify the text, sometimes to its detriment.

Another serious challenge when reading alchemical descriptions of processes is the identification of ingredients. Defining the materials used in a process is frequently very difficult, since most of the words used to designate them are polysemic and may refer to various substances, depending on the period, the place, and even the context. But even when we know to what material a term refers, this material was usually impure in the Middle Ages. Inevitably, this often leads to the impossibility of knowing precisely the ingredients and the resulting material of medieval processes, and thus to conjecture. Furthermore, the impurities of the ingredients may sometimes play a role in the processes (Principe 1987). This problem of identification is intensified by the alchemists' habit of using codenames and allegorical terms: if some of them are very frequent and easy to solve, such as "sun" (*shams/sol*) for gold, others are rather complicated to understand, such as "nosebleed" (*ruʿāb*) for copper (Siggel 1951: 40). Allegorical names are also sometimes used for processes, such as the alchemical "marriage" (*tazwīj/coniugium*) for the final combination of the ingredients in the alchemical work.

The way in which many of these treatises were composed also raises difficulties. Indeed, alchemists copied ancient treatises and borrowed from many compilations of recipes, sometimes by altering the processes or modifying the ingredients. And the formula "*mujarrab*"/"*probatum est*" (it has been proved, tested) did not guarantee that the recipe as it has been written had really been tested.

No comprehensive study has been published on the topic of medieval alchemical processes in general, but many researchers have studied medieval processes in their publications by as early as the late nineteenth century and early twentieth century, such as Marcellin Berthelot, Julius Ruska, or Henry Stapleton. More recently, a few publications have dealt with specific processes, but never with the goal of a global synthesis on the topic (e.g. for antiquity, see Martelli 2009; and for the Middle Ages, see Newman 2000; Moureau 2016a; Moureau 2016b; Moureau and Thomas 2016).

The sources for studying alchemical processes are of four kinds: textual, archaeological, iconographic, and experimental. Textual sources are pivotal for the understanding of processes and are by far the principal way of knowing them today. Although they are usually full of tacit and implicit knowledge, they are in most cases the only way to know how a process was conducted. Archaeological sources can also help, but to a much lesser extent than for alchemical equipment (see Chapter 3). Analysis of medieval tools and materials sometimes brings to light unknown elements on the processes. Iconographic sources also provide us with valuable information additional to textual material. There is a fourth kind of source, unfortunately less known and exploited nowadays: experimentation. Reproducing medieval processes allows the researcher to shed new light on them (Fors et al. 2016; see also Moureau and Thomas 2016; Principe 2016; Rampling 2018).

A categorization of the processes can help to describe them. Yet, processes are hard to classify. Indeed, the huge number of parameters that one needs to take into account makes all classifications imperfect. In addition to this, the significance of interpretation when studying alchemical processes may also lead to bias. In medieval classifications, alchemists usually order processes according to their temporal succession in the alchemical work; and since the sequence of the alchemical work often varied among alchemists, the same goes for classifications. However, modern categorizations are so distant from medieval thought that it would be inappropriate to use them in this chapter. This is why the next part of this chapter is a description of two major classifications in medieval alchemy: the first one, taken from the Arabo-Muslim literature of the second to third centuries AH/eighth to ninth centuries CE, is the classification that is found in Abū Bakr al-Rāzī's *Kitāb al-asrār* (*Book of Secrets*), a technical treatise that can be considered as one of Rāzī's *opera magna* (edited in Dānish-Pazhūh 1964; partially translated into English in Stapleton et al. 1927; translated into German in Ruska 1937); the second one is taken from one of the most influential Latin medieval texts, the *Summa perfectionis*, a pseudepigraphic treatise attributed to Geber (the Latin name of Jābir b. Ḥayyān), most likely written at the end of the thirteenth century, possibly by Paul of Tarento (edited and translated into English in Newman 1991). The last part of this chapter consists of a more detailed description of the most significant alchemical processes.

PROCESSES IN RĀZĪ'S *KITĀB AL-ASRĀR* AND PSEUDO-GEBER'S *SUMMA PERFECTIONIS*

Rāzī's *Kitāb al-asrār* is entirely arranged according to processes. After an introduction on instruments and materials, the text is divided into seven chapters, each one dedicated to processes or sections of the alchemical work.[2] The description below explains the general structure according to one of the alchemical works of Rāzī; various smaller recipes for making gold or silver from various base metals are inserted into different chapters. The alchemical work in Rāzī, as in most Arabic alchemical treatises, is made up of three main ingredients: a body (i.e. a base metal such as lead or copper); a spirit (i.e. a substance that can be sublimated, usually mercury); and a soul (i.e. an elixir). Rāzī follows the Jābirian elixir theory (see Chapter 1). He also provides the reader with other recipes for chrysopoeia and argyropoeia.

Rāzī first addresses the question of the preparation of spirits. The most prominent spirit is mercury, and its preparation is either by coagulation or by sublimation. Then, Rāzī moves on to the preparation of the body, namely the base metal, and begins with calcination (i.e. the reduction of the metal into a powder; Dānish-Pazhūh 1964: 32–56).[3] Afterwards, he explains how ceration works (Dānish-Pazhūh 1964: 56–75): in this process, the calcinated metal is given the consistency of wax. Dissolution is the next process, in which alchemists make the cerated metal become fluid (Dānish-Pazhūh 1964: 76–83). After a short section on how to mix the dissolved components, namely the dissolved body (i.e. dissolved metal), the dissolved spirit (i.e. cerated and dissolved mercury), and the dissolved soul (i.e. the elixir in all likelihood), namely the union of the three components of the alchemical work (Dānish-Pazhūh 1964: 83–4), Rāzī focuses on coagulation (*'aqd*), which is making solid a liquid material (Dānish-Pazhūh 1964: 84–6). The penultimate chapter is devoted to sublimation (Dānish-Pazhūh 1964: 86–8). The last section of Rāzī's book concerns "reddening waters," namely tinctures and elixirs (Dānish-Pazhūh 1964: 90–116). In this section, the author explains how to make solutions that can dye metals, made from various materials (vitriols, sulfur, iron rust, etc.). Among these tinctures, Rāzī also broaches the elixirs, which are made by distillation *per ascensum* (see the description in the next section below). The making of elixirs has been explained in Chapter 1.

The *Summa perfectionis* is one of the most influential alchemical Latin texts, and its fame is maybe due to its unusually clear and systematic approach. In this text, the author offers a synthetic and detailed description of alchemy. He proposes a different alchemical work from that of Rāzī. Instead of using the Jābirian elixir theory, in which organic materials are often used as a base material for the elixir, he only uses mercury for the creation of the tincture (*medicina*, "medication" in the text). He also follows a different order for the operations. The author of the *Summa perfectionis* organizes his writing in a completely different way to Rāzī. The first part of the first book of the *Summa perfectionis* is dedicated to theoretical discussion on alchemy, taking part in the lively debate over the possibility of transmutation in the Latin world; in the second part of the first book, the author focuses on materials and describes them. The second book contains the general description of the alchemical work and of the processes. This is the book that will be mostly used in this chapter. The final book deals with particular descriptions of processes; it is more practical, and finishes with a discourse on alchemical "medicines" (for metals, not for humans), the description of several processes, and the end of the alchemical work.

When explaining processes theoretically, the author first presents sublimation (*Summa perfectionis* = *SP*, *cap.* 39), and then turns to "descension" (*descensio*; *SP*, *cap.* 45) – a way of purifying materials – and distillation (*SP*, *cap.* 46). There follow in order calcination (*SP*, *cap.* 47), dissolution (*SP*, *cap.* 48), coagulation (*SP*, *cap.* 49), fixation (*SP*, *cap.* 50), and ceration (*SP*, *cap.* 51), all of which will be described in the next part of this chapter.

The third book gives interesting descriptions of processes that are not specifically linked with alchemy (hence their place in the book), but that are more secondary techniques than steps of the alchemical work: some of them can be used in the operations that are described in the second book. The author of the *Summa perfectionis* broaches the subject of firing (*ignitio*) and fusion (*fusio*), where he gives advice on how to heat and melt metals (checking their colors, etc.; *SP*, *capp.* 85–6). Then he mentions the exposure of metals to "acute" waters, namely corroding vapors (*SP*, *cap.* 87), such as making white lead (a lead carbonate, $2PbCO_3{\cdot}Pb[OH]_2$) or verdigris (a copper carbonate, $Cu_2CO_3[OH]_2$), by suspending metal strips over vinegar. Finally, he also explains "extinction" (*extinctio*), namely the tempering of metal (*SP*, *cap.*

88). In this part, we also find two metal purification operations: cupellation and cementation, the last one not to be confused with modern cementation in metallurgy, such as iron carburizing (to make steel) or copper cementation with zinc to make brass (*SP*, *capp.* 82–4).

THE MAIN ALCHEMICAL OPERATIONS

The most famous nonmetallurgic operation used in alchemy was *distillation.* There were two kinds of distillation: distillation *per ascensum* (i.e. by raising the material) and distillation *per descensum* (i.e. by bringing it down). These two operations were clearly distinguished. They were both abundantly used also by craftsmen and particularly by apothecaries (Thomas 2009; Thomas and Claude 2011).

Distillation *per ascensum* was, in the Middle Ages, the process in which a substance, most often a liquid, was volatilized and then condensed into a liquid form. The main aim of distillation was to divide a material into its components, hence also sometimes to purify substances. This was done by using an alembic with a spout (see Chapter 3). The material was placed in the cucurbit, the head of the alembic was set on the cucurbit, and the joint was luted (see Figure 2.1). A receptacle was placed at the end of the spout. Then the apparatus was gently heated in a water bath (for lower temperatures, since water boils at 100°C) or a pot full of ashes (for higher temperatures; i.e. a thermal insulator, in order to control the temperature more easily and avoid thermal shocks). The material vaporized and rose into the head, where it condensed, since the head was colder than the cucurbit (sometimes the head was cooled down with a wet sponge from the outside). The condensed material flows into the collecting channel of the head, then into the spout, and finally into the receptacle. Components of the material distilled out according to temperature, one after the other. Distillation *per ascensum* in alchemy was mostly used to isolate the four elements of the "stone" when making the elixir (see Chapter 1). This is particularly true in Arabic alchemy and in early Latin alchemy, in texts indebted to Arabic theories. In later Latin theories, as in the *Summa perfectionis*, distillation was generally used to purify substances. From the time of the appearance of the quintessence theory (see Chapter 1), distillation became a central operation again in order to produce alcohol (Thomas 2020). In the Arabo-Muslim world, distillation *per ascensum* was very common and most often used in rosewater workshops. A detailed description of it can be read in the *Nukhbat al-dahr* (*Selected Pieces of Time*) written by Shams al-dīn al-Dimashqī in the early eighth century AH/ fourteenth century CE (*Nukhbat al-dahr*, *faṣl* 9, *waṣf Falisṭīn wa-al-Urdun*, in Mehren and Fraehn 1866: 194–8). In the Latin West, distillation was very popular for the manufacture of both alcohol and rosewater, but also for purifying acids or to obtain incendiary products (on incendiary products, see, for instance, Marcus Graecus' *Liber ignium* in Berthelot et al. 1893: 1:100–20).

In the West, two other operations were derived from distillation *per ascensum*: *rectification* (from the Latin word *rectificare*) and *cohobation* or *circulation* (Latin *cohobare*). To rectify means to distill the distillate several times (see Figure 2.1).[4] The separation of the substances contained in a liquid to be distilled is based on different boiling and vaporization temperatures. During distillation, the separation is not complete: the vapors still contain quantities of the two liquids that are to be separated. In the distillate, materials that have a lower boiling point are found in higher proportions than materials that have a higher boiling point. By repeated distillations, the process thus allows the purity of the distillate to increase little by little. We know that this technique was applied in the production of alcohol, but not exclusively. For instance, the author of the *Summa perfectionis* recommends repeating the operation in the case of water distillation in order to obtain pure water (Newman 1991: 409). The physician Theodoric de Cervia distinguished between *aqua perfecta*, produced by a series of seven distillations, and *aqua perfectissima*, which required ten distillations, and that some scholars consider to have possibly been 90° alcohol (Halleux 1981: 249). The cohobation or circulation was a repeated

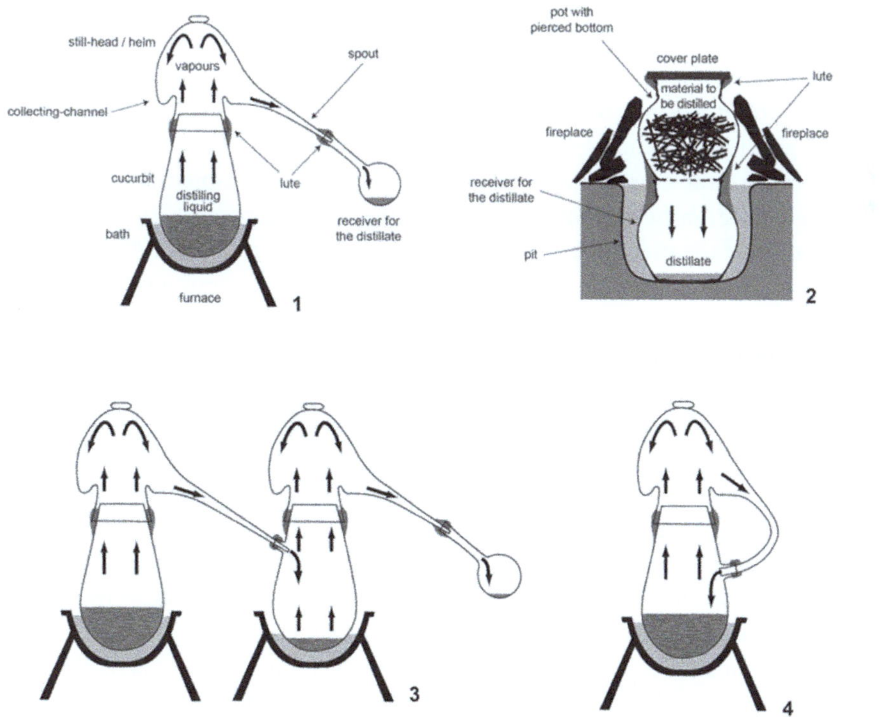

FIGURE 2.1 Schematic representations of alchemical processes. 1. Distillation *per ascensum*. 2. Distillation *per descensum*. 3. Rectification (repeated distillation). 4. Cohobation or circulation (circular distillation). Drawing by Nicolas Thomas.

distillation of the same material, with the distillate falling back into the material to be distilled in the bottom of the cucurbit (see Figure 2.1). It was used in particular by John of Rupescissa for producing the quintessence (Halleux 1981: 254). In modern terms, it corresponds to the process of refluxing, rather than being a distillation; the product can be altered by being maintained at a certain temperature for very long periods, but there is no separation of a material.

Distillation *per descensum* was a different process (see Figure 2.1). It must not be confused with the descension of metals using the descensory or *botus barbatus* (see below), although the device, also called *descensorium* or *chimina* in the *Summa perfectionis*, looks very similar (see Chapter 3). The descensory is a pot with a perforated bottom placed on top of another pot. Vegetable material is placed in the upper pot of the descensory, then this upper pot is sealed and gently heated. The oil contained in the vegetable substance flows into the lower pot (Thomas and Claude 2011). This process was not used frequently in alchemy, but the author of the *Summa perfectionis* mentioned it as a way to extract combustible oils from vegetable materials, which would be destroyed in distillation *per ascensum* (*SP*, *cap.* 46).

As a final distillation, the author of the *Summa perfectionis* mentions filtration (*SP*, *cap.* 46). Indeed, in its original meaning, distillation (the Arabic *QṬR* and the Latin *distillare*) means "to flow in drops," hence this classification. The distillation *per filtrum* allowed the alchemist to extract substances without fire.

Descension was a process that consisted of making a metal drop down and filtering it in a specific device. It was called *istinzāl* in Arabic and *descensio* in Latin. In Arabic as in Latin texts, the operation was sometimes designated by the device, *botus barbatus* (on this name, see Chapter 3), such as in the translation of the pseudo-Avicennian *De anima* (Moureau 2016b: 494, 727; see also Berthelot et al. 1893: 1:162). The most detailed description of this process is found in the *Summa perfectionis* (*SP*, *cap.* 45).[5] The descensory consisted of a crucible pierced in its bottom and placed on top of another crucible (see Figure 2.2). The device is very similar to the one used in distillation *per descensum* (see above), which is why the two operations are often confused. Alchemists clearly distinguished between the two processes: the temperature is high for descension, low for distillation.[6] The joint was luted and a lid might be placed on the upper crucible if needed. The upper crucible was heated, and the metal flew through the holes into the lower crucible. The author of the *Summa perfectionis* gives three uses for this process. The first reason for using the descensory is to confirm that a metal can be melted. Secondly, descension allows one to avoid losing too much metal in the calcination process; indeed, since calcination of a metal is a process that must be repeated (not all the metal is calcinated at once), when calcinating a metal that is easily corroded, the descensory allows the metal that has not been calcined to flow into the lower crucible and to be protected from the fire that would corrode it. Finally, descension is a very common way of purifying metals, since the impurities stay in the higher crucible while the metal goes into the lower crucible.

Calcination (*taklīs* in Arabic) in the Middle Ages is different from the present definition of calcination: it was the process of reducing a metal into a calx (*kils/calx*), namely a powder. The name of the process itself comes from the Latin term *calcinatio* (i.e. *calcem facere*, "to make a calx"). The idea behind this process in the mind of medieval scholars was to reduce the metal into extremely fine particles. In the *Summa perfectionis*, the author explains that the pulverization is made by the "removal of the humidity consolidating its particles," and that the aim of the process is to burn the "burnable sulfureity which corrupts and soils" (*SP*, *cap.* 47). The processes usually involved sulfur compounds: in fact, the calx consisted mainly of metal sulfides (more rarely, metal oxides). In most cases, calcination is what is called today corrosion. Calcination is carried out in general by cementation (see below). The entire metal is rarely calcinated in one calcination, and the process must be repeated in order to calcinate all the metal. The temperature for the process needs to be high. Gold is an exception: since it does not easily mix with sulfur or oxygen, gold was usually calcinated by adding lead to it and then crushing the alloy, because lead makes gold brittle.[7] Calcination can also be applied to stones and salts.

The name of *ceration* comes from the Latin *ceratio* or *inceratio*, from *cera*, "wax," a translation from the Arabic *tashmī'*, from *sham'*, "wax," since its purpose is to give to the calcinated metal the consistency of wax. Indeed, after reducing the metal into tiny particles, Rāzī wanted to have the metal in the form of a soft malleable material. The author of the *Summa perfectionis* adds that ceration is the liquefaction of a hard material that is not fusible. The chemical and physical processes involved in ceration are very hard to identify, since they are a series of complicated mixtures with various materials (spirits, salts, boraces): ceration actually refers to a wide range of operations in modern terms and is a conspicuous example of the difference between medieval and modern chemistry, since "making something like wax," according to the medieval definition, is, from the modern point of view, a series of various processes depending on the materials used and the recipes. Ceration could be done in various kinds of containers, usually sealed with lute. The temperature for the process needed to be moderate. Ceration could also be applied to spirits, salts, and stones.

Dissolution (*taḥlīl/dissolutio*) is the operation by which alchemists wanted to make a dry substance (the cerated metal or sometimes directly the calx in Rāzī's book) become liquid by dissolving it in corrosive agents (literally "sharp waters," *miyāh ḥādda/aquae acutae*). They used a wide range of corrosive agents, from vinegar to soda. Here, again, it is in most cases impossible to identify the chemical and physical processes when they were applied to the cerated metals (which is also hard to identify), and each recipe needs to be investigated separately. Dissolution was conducted with a very moderate temperature, a gently fired water bath, or fermenting dung. Dissolution could also be applied to spirits, boraces, salts, and various other materials.

Coagulation (*ʿaqd, taʿqīd/coagulatio, congelatio*) was the process by which alchemists made a liquid material become solid by means of "the removal of what is humid" (*SP*, *cap.* 49). They applied this term to any kind of coagulation, from milk coagulating into cheese to water congealing into ice, as well as the coagulation of mercury. According to Rāzī, there are four kinds of coagulation, which correspond to different recipes, but all produce a metal that is the solidification of the three dissolved ingredients (body, spirit, and soul; see above). It is considered the end of the alchemical work by Rāzī (Dānish-Pazhūh 1964: 13). For the author of the *Summa perfectionis*, there are two aims of coagulation: hardening mercury and "removing from dissolved medicines the wateriness of mercury that is mixed with them" (*SP*, *cap.* 49). Coagulation of mercury is almost an obsession in alchemical texts. Alchemists wanted to make solid mercury, which is fluid at room temperature. For this, they usually mixed mercury with another metal, in most cases lead or tin, which created a solid amalgam. A fairly common recipe for coagulating mercury deserves special mention (see, for an example, Moureau 2016b: 122, n. 234). To coagulate mercury, it was often placed in an iron pot (iron does not amalgamate with mercury), which was placed above another pot containing molten lead and tin. The lead and tin vapors amalgamated with the mercury and hardened it. This recipe, a rather complex one for a process that is actually quite simple (mercury could have simply been mixed with hot lead or tin), often puzzles the modern reader. The use of metal vapors may have been a way of avoiding the loss of mercury by vaporization (mercury readily vaporizes when heated).

The *Summa perfectionis* mentions a process that is not found in Rāzī's book: *fixation* (*fixio*; *SP*, *cap.* 50). Fixation was the "agreeing adaptation" (*conveniens adaptatio*) of a thing that flees fire (i.e. a volatile substance); its aim was to fix a change in the properties of a substance. This operation is not often systematically explained as in the *Summa perfectionis*; since it is a kind of super-operation, depending on the material that needed to be fixed, fixation was carried out by various processes. For instance, metals were fixed by calcination, while sulfur and arsenic were fixed by repeated sublimation. So, fixation was not a process in itself but an operation that could be carried out using various processes.

Sublimation (*taṣʿīd/sublimatio*) was, in the Middle Ages, the process that consisted of volatilizing a substance and condensing it in a solid form (see Figure 2.2). It was used for various purposes. One was to purify impure materials, such as the sublimation of sulfur, arsenic (i.e. arsenic sulfides, As_2S_3 and As_4S_4), etc. Another aim was the production of compounds, such as the making of cinnabar (HgS) from mercury. Sublimation of mercury was one of the most important processes for alchemists. In this operation, they usually mixed mercury with common salt (i.e. a compound of chlorine) and vitriol or alum, whose decomposition produced sulfuric acid, acting as an oxidizing

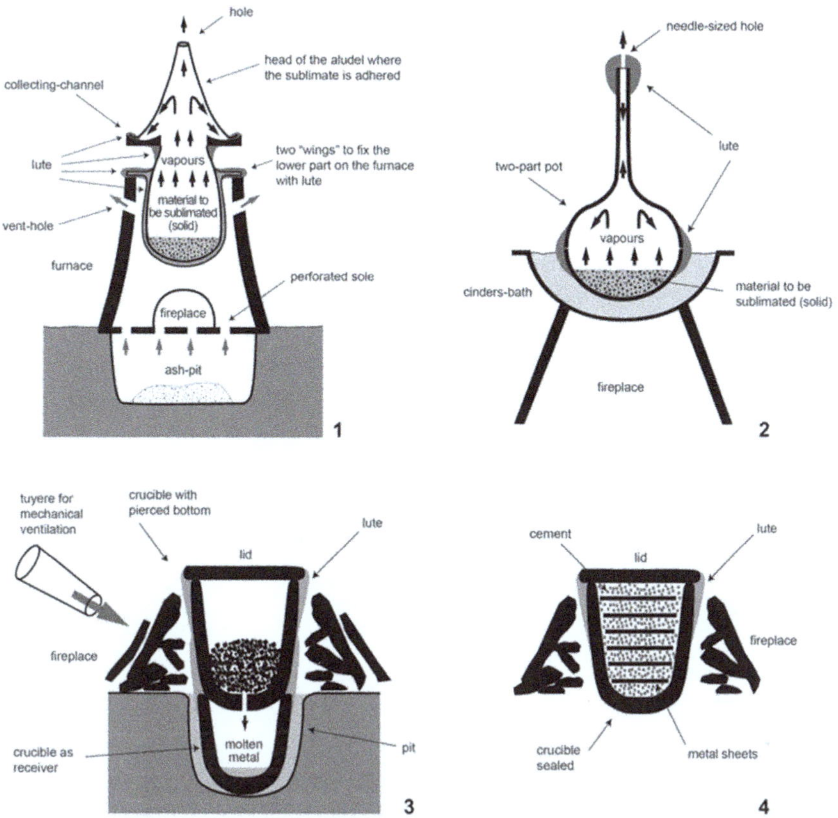

FIGURE 2.2 Schematic representations of alchemical processes. 1 & 2. Sublimation. 3. Descension. 4. Refining by cementation. Drawing by Nicolas Thomas.

agent. And then they sublimated the mixture: the process produced sublimated mercury, better known under the name of corrosive sublimate ($HgCl_2$), a strong poison that could be used to dye metals, since it easily decomposes and creates a surface alloy.[8] Another important sublimation of mercury was the production of cinnabar by mixing mercury and sulfur and sublimating them. Sublimation could be carried out in various devices (called an "aludel"), from very simple bottles to complicated apparatus (on these instruments, see Chapter 3). The process went as follows: the material was placed in the bottle or, in the case of a complex aludel, in the lower part of the aludel (i.e. a large pot covered with a top part sealed with lute). The aludel was placed on a furnace, usually a natural cylindrical draught stove (*mustawqad* in Arabic), and then heated. The material sublimed and solidified on the higher part of the aludel (the neck, if the aludel is a bottle, or the lid, if the aludel is more elaborate); then the alchemists gently scraped off the sublimate to collect it.

Cupellation (there was no specific word for this in Arabic, usually simply *taṣfiya*, "purification"/*cupellatio* or *cineritium*, from *cinis*, "ashes," since it involves ashes) is a process by which silver or gold are purified from other metals alloyed with them (Moureau and Thomas 2016; Saussus and Thomas 2019) (see Figure 2.3). This process was most often used for the extraction of silver from lead ore, or for assaying, particularly in the context of minting coins. The principle of cupellation is that silver and gold do not corrode or burn when heated (in modern terms, they do not oxidize easily when heated), while other metals known in that period do. Lead is mixed with silver or gold, and the whole is placed in a cupel, namely a crucible made from ashes (or a crucible filled with ashes), and then heated. A flow of oxygen is projected onto the metal (by means of a bellows and a tuyere or by convection), which oxidizes the alloyed metals, and the oxides flow into the ashes (the porosity of ashes allows the oxides to penetrate, but not the molten metals) and leaves the silver or gold alone at the top.

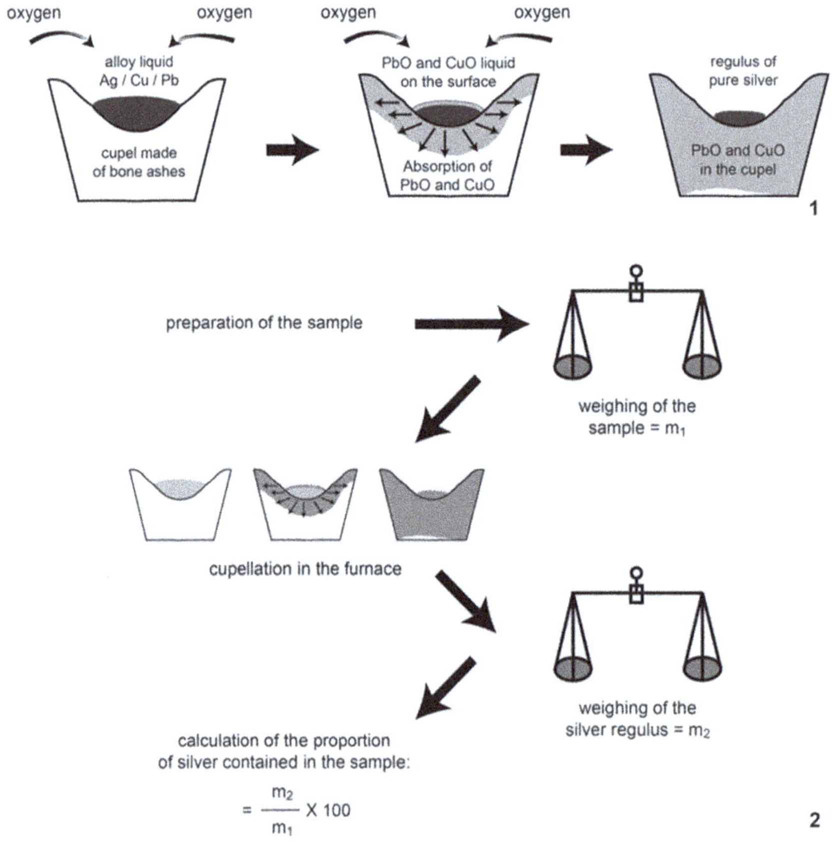

FIGURE 2.3 1. Schematic representation of cupellation. 2. Schematic representation of fire assay. Drawing by Nicolas Thomas.

The following technique described in the *Summa perfectionis* is *cementation* (no specific word in Arabic/*cementum* or *caementum*; see Figure 2.2; Ramage and Craddock 2000; Thomas 2006; Celauro et al. 2017). Cementation was mostly used by alchemists to purify metals. The *cementum* was a powder made up of the materials (depending on the recipe, it could be urine with copper sulfate and powdered bricks or salt) that the alchemist wanted to react with the nonprecious metal alloyed with gold or silver, which was placed in contact with the alloy and then heated, always below the melting temperature of the metal. The cement and the metal, cut into small sheets, were placed in a crucible, layer over layer.[9] The whole was heated, and the nonprecious metal reacted with the other ingredients to form compounds such as copper chloride or other oxides. This is a surface reaction that takes a long time, usually twenty-four hours.

CONCLUSION

The most striking observation when looking at alchemical operations is the continuity and stability of processes throughout medieval history. The penetration of alchemy into the Arabo-Muslim world was not a revolution of processes, most of which were already known, but rather a development and improvement. The transition to the Latin world, on the other hand, was more striking: the Latins discovered several techniques previously unknown in the West. After assimilating them, they in turn developed and improved them, such as distillation (with rectification or cooling coils). Moreover, many of these operations continue to be used in modern times, but with a difference in scale and the beginning of mass production, particularly for alcohol from the late fifteenth and early sixteenth centuries.

However, this stability must not give the impression of inert knowledge. Processes go far beyond the borders of alchemy and circulated through a large number of fields of craftsmanship, from metallurgy to pigment manufacturing. These fields exchanged and appropriated techniques and readily borrowed processes from each other.

CHAPTER THREE

Laboratories and Technology: *Alchemical Equipment in the Middle Ages*

NICOLAS THOMAS AND SÉBASTIEN MOUREAU

INTRODUCTION

A laboratory is first and foremost a place (i.e. a space limited by walls). For the medieval period, laboratories where alchemy was practiced are poorly known, since we have very few examples.[1] However, a laboratory is also a place where tools were used. Unlike laboratories, modern researchers can study this equipment from various sources and even imagine the alchemist at work.

The tools, instruments, and apparatus used by alchemists to conduct their operations were of various kinds, from simple hammers to more complex devices. Tools and laboratories are objects for which the line between alchemy and craft is very hard to draw. Indeed, while the aims of alchemists and craftsmen were different, as explained in Chapter 2, the tools were often the same for both communities.[2] Therefore, one can define an alchemical tool as any instrument that was used by alchemists for their operation, and not only the apparatus specific to alchemists.

A few studies have been published on alchemical equipment in the Arabo-Muslim world, but they are old, and further inquiries would be necessary to update them. Stapleton and Azo wrote an article on this in 1905 and with Ḥidāyat Ḥusayn in 1927 (Stapleton et al. 1927), where they focused on Arabic texts describing apparatus. Still for the Arab world, Eilhard Wiedemann studied alchemical instruments in 1909, and, in 1923, Julius Ruska published a brief note on the topic. In 1956, Holmyard wrote a short synthesis on alchemical equipment, but not specifically focused on the Middle Ages.

Other papers have dealt with particular apparatus, such as Greek alchemical tools in Zosimos' works (Mertens 1995: 113–61; Martelli 2011) or various type of glass equipment (Savage-Smith 1997; Kurzmann 2009). The question of the aludel (sublimation device) was revisited recently (Moureau and Thomas 2015). For the medieval West, other studies have been devoted to operations, such as distillation (Taylor 1945; Forbes 1970; Moorhouse et al. 1972; Thomas 2009) or distillation *per descensum* (Moorhouse 1981; Thomas and Claude 2011). Some publications of archaeological discoveries of alchemical devices complete these studies (Rouaze 1986; Kamber et al. 1999). In addition, a good approach to alchemical iconography was proposed by van Lennep (1984).

SOURCES

The sources available for the study of alchemical equipment are diverse and varied. They are of four kinds: textual sources, iconographic material, archaeological remains, and also experiments.

Textual sources

Texts dealing with alchemical devices can themselves be divided into different types. The most frequent accounts are small descriptions inserted into recipes; most of time these are very elliptical and are usually intended for readers who are familiar with the practice. It is impossible to list them all, and they rarely contain complete descriptions of instruments. Some texts provide lists of devices, which vary in length. Interestingly, they provide us with a synthetic and global picture of the alchemists' equipment, as well as the vocabulary used. These lists are sometimes accompanied by very short descriptions, such as, for Arabic, in the *Mafātīḥ al-'ulūm* (*Keys of Wisdom*), a handbook addressed to secretaries (*kuttāb*) written between 366 AH/976 CE and 387 AH/997 CE by Abū 'Abd Allāh Muḥammad b. Aḥmad al-Kātib al-Khwārizmī a state officer at the court of the Sāmānid ruler Nūḥ II b. Manṣūr, in which a complete chapter is devoted to alchemy (edited in Vloten 1968: 255–8; translated into English in Stapleton et al. 1927: 362–3); the *Kitāb al-ḥabīb* (*Book of the Beloved*), an early Arabic alchemical text (edited and translated into French in Berthelot et al. 1893: vol. 3, 35–6, 77); and for Latin, in the *Speculum naturale* and *Speculum doctrinale*, finished in 1259 by Vincent of Beauvais, an encyclopedist who gave a large place to alchemy in his work (Moureau 2012: 49). However, these inventories do not provide substantial information or detailed accounts on the devices.

Much more interesting are the longer descriptions that are found in several texts. Although they are not frequent, they raise elements of the highest importance for the understanding of the alchemical equipment. A few texts contain long accounts on the topic in the Arabo-Muslim world. Among them are the following. The Jābirian corpus contains descriptions of instruments scattered throughout the texts that compose it. However, these descriptions often are not very systematic, and their dispersal across multiple treatises makes them difficult to follow (Kraus 1942: 11, n. 1).

Abū Bakr al-Rāzī's *Madkhal ta'līmī* (*Instructive Introduction*) and *Kitāb al-asrār* (*Book of Secrets*) are practical and technical texts of the third to fourth centuries AH/ninth to tenth centuries CE, where Rāzī provides the reader with descriptive lists of instruments and instructions on how to build them (*Kitāb al-asrār* edited in Dānish-Pazhūh 1964; partially translated into English in Stapleton et al. 1927; translated into German in Ruska 1937; *Madkhal ta'līmī* edited in and translated into English in Stapleton et al. 1927: 412–17, 345–61).

The *'Ayn al-ṣan'a wa-'awn al-ṣan'a* (*Source of the Art and Help of the Art*) was written in Baghdad by Abū al-Ḥakīm Muḥammad ibn 'Abd al-malik al-Ṣāliḥī al-Khwārizmī al-Kāthī in 426 AH/1034 CE (Stapleton and Azo 1905). This technical text, deeply indebted to Rāzī's works, includes a few descriptions of alchemical instruments. It has also been translated into Persian (Maqbūl 1929).

The pseudo-Avicennian *De anima* is the Latin translation and compilation of three Arabic texts that have not been preserved (edited and translated into French in Moureau 2016a; Moureau 2016b). The longest part of the text was written between the third quarter of the eleventh century and the middle of the thirteenth century, and may have been translated around 1226 or 1235. This complicated text contains descriptions of the manufacture of various instruments.

In the Latin West, in addition to the pseudo-Avicennian *De anima*, we also find a few texts with long descriptions, among which we may highlight the following. The *Liber secretorum Bubacaris* is a translation, or more precisely a paraphrase, of Abū Bakr al-Rāzī's *Kitāb al-asrār* (studied and partially edited in Ruska 1935). Although much less clear than the Arabic version, this text presents descriptions of instruments.

The *Summa perfectionis* of Pseudo-Geber is by far the most comprehensive text on the manufacture of alchemical apparatus in the Middle Ages (edited and translated into English in Newman 1991). This pseudepigraph attributed to Geber (Latin name of Jābir b. Ḥayyān) was written at the end of the thirteenth century, perhaps by the monk Paul of Tarento. Its systematic approach includes precious descriptions on how to make instruments.

A *Theorica et practica*, another treatise on alchemy that can for its part be attributed with certainty to Paul of Taranto, offers a good overview of the apparatus used in the thirteenth century (edited in Newman 1986: vol. 3).

The text attributed to Albert the Great entitled *Calisthenes* (from its incipit) provides a precise description of how to make an aludel (edited in Kibre 1944).

There are also descriptions in other languages, such as the *Petit Rosaire de maistre Arnault de Ville Nove sur la Rose fait et composé d'alquimie*, in a manuscript kept at the library of Cambrai and written in 1426 (MS 918). Rather than a translation of the *Parvum rosarium* attributed to Arnald of Villanova, as one would expect from the title, it is a very free and distant adaptation of the *Rosarius philosophorum*.[3] In this manuscript, one finds descriptions in French of the making of alchemical instruments, as well as numerous images of apparatus.

Another textual source category is craft treatises, since most alchemical instruments were identical to those used by craftsmen. In Arabic, one may cite al-Ḥasan b. Aḥmad al-Hamdānī's *Kitāb al-jawharatayn* (*Book of the Two Substances*), a treatise of the fourth century AH/tenth century CE on mining and metallurgy (edited and translated into German in Toll 1968; Moureau and Thomas 2016), or chapter 9 of the *Nukhbat al-dahr* (*Excerpts of Time*), written in the early eighth century AH/fourteenth century CE by Shams al-dīn al-Dimashqī on the distillation of rosewater (*faṣl* 9, *waṣf Falisṭīn wa-al-Urdun* edited in Mehren and Fraehn 1866: 194–8). The same applies to Latin texts, such as the *Diversarum artium schedula* written around 1125 by the monk Theophilus (perhaps Roger of Helmarshausen), in which several descriptions are found in recipes, whether for the making of pigments or for metallurgy (edited and translated into English in Dodwell 1986; other editions in Ilg 1874; with a French translation in L'Escalopier 1843; translated into English in Hawthorne and Smith 1979; translated into German in Brepohl 1999).

Since alchemy was close to medicine in the Latin West in its quest to limit bodily corruption, particularly from the thirteenth century onwards, medical or pharmaceutical treatises provide numerous descriptions of apparatus, especially for distillation. Some of these accounts are in vernacular languages.[4]

Finally, references in accounts or inventories are rarer and more difficult to find, but no less interesting. These mentions mainly concern alembics for the production of rosewater or *aqua ardens* (alcohol; Bénézet 1999; Thomas 2009). Although these sources do not provide detailed descriptions, they do provide information on the social context of the use of the apparatus, the materials most commonly used in their manufacture, or an assessment of their value.

Iconographic sources

Unfortunately, medieval Arabic alchemical manuscripts contain very few images. Only in more recent manuscripts do illustrations appear. These include alchemical instruments in MS Rabāṭ, Khizāna Ḥasaniyya, 1393, fol. 153r (aludels and alembic), MS Rampur, Raza Library, Kīmiyā' 12, fol. 132v–3v (ninth century AH/fifteenth century CE; pots and aludels), and MS Bethesda, National Library of Medicine, A65, fols 80v–1v, 83v–4r (1123 AH/1712 CE), but these examples are uncommon.[5]

Two books, however, are noteworthy for their exceptional nature. The *Muṣḥaf al-ṣuwar* (*Tome of Images*) is an Arabic treatise attributed to Zosimos of Panopolis that consists of a text illustrated by enigmatic images. These pictures are allegorical, but in one of them an alembic is pictured at the top of a great furnace (fol. 153r). The treatise is preserved in the single manuscript Istanbul, Arkeoloji Müzeleri Kütüphanesi, 1574 (668 AH/1270 CE; facsimile in Abt 2007b; English translation in Abt and Fuad 2011, to be read with the remarks in Hallum 2009).

Abū al-Qāsim al-ʿIrāqī's *Aqālīm sabʿa* (*Seven Climates*) is an alchemical text illustrated with numerous images. Most of them are allegorical (from human figures to hieroglyphs), but there are also a few alchemical instruments depicted in the text (not edited; for this study, we have used MS London, British Library, Add. 25724 of the twelfth century AH/eighteenth century CE).

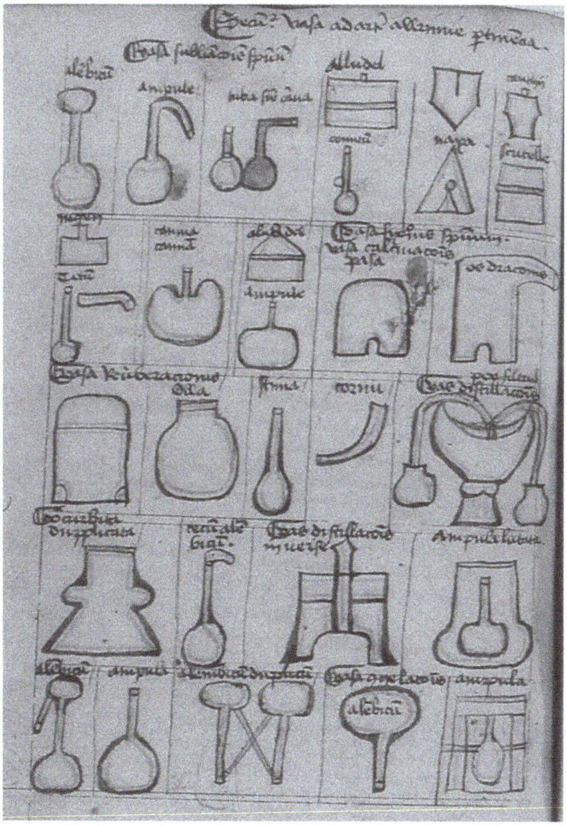

FIGURE 3.1 Alchemical apparatus. Bibliothèque nationale de France, Paris, MS Lat. 7162 (fifteenth century), fol. 92v.

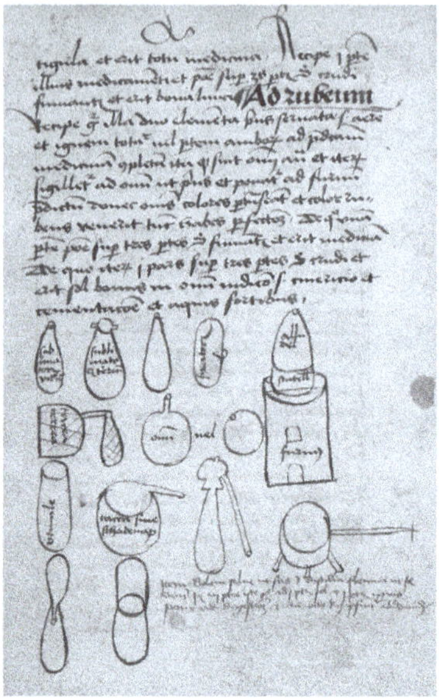

FIGURE 3.2 Aludels, alembic, furnace, and other alchemical equipment. University of Pennsylvania Library, Philadelphia, Edgar Fahs Smith Mem. Coll., MS Codex 69 (ca. 1450–75), fol. 22v.

FIGURE 3.3 On the left, fractional distillation. On the right, cohobation or circulation. *Ordinal of Alchemy* of Thomas Norton, British Library, London, Add. MS 10302 (1477), fol. 37v.

On the other hand, iconographic material is abundant in Latin sources (Figures 3.1–3.3). Sketches of apparatus appear in many manuscripts, such as MS Paris, Bibliothèque nationale de France, lat. 6514 (fourteenth century), fols 68r–72r, MS Paris, Bibliothèque nationale de France, lat. 7156 (fourteenth century), fols 139r–41v, MS Jena, Thüringer Universitäts- und Landesbibliothek, El. q. 18 (early fourteenth century), fol. 60v, MS Cambridge, Trinity College, O.2.18

(fourteenth to fifteenth centuries), MS London, British Library, Harley 2407 (1450–1500), fols 106v–11, MS Philadelphia, University of Pennsylvania Library, Edgar Fahs Smith Mem. Coll., Codex 69 (ca. 1450–75), fol. 22v, fol. 120v, MS London, British Library, Add 10302 (1471), MSS München, Bayerische Staatsbibliothek, Clm 405 (fourteenth to fifteenth centuries), fols 53r, 157v, 169r, 171r–v, 176v and Clm 25110 (fifteenth century; images available in Ganzenmüller 1941), and MS Paris, Bibliothèque nationale de France, lat. 7162 (fifteenth century), fols 92v–3r. Printed works are even more bountiful from the end of the fifteenth century onwards.[6]

Medieval manuscripts in other languages contain illustrations of alchemical equipment as well, such as, for Greek, MS Venice, Biblioteca Marciana, Gr. Z. 299 (tenth to eleventh centuries), fols 188v, 193v, 194v, 195v–6v, or, for French, MS Cambrai, 918, mentioned above, MS Kassel 4° med. 1 (fifteenth century), especially fols 324r–6r, or MS London, Wellcome Library, Western 446 (fifteenth century), which contain the *Ymage de vie*, an alchemical treatise with an abundant and well-ordered imagery (Neven 2014; facsimile and study in Dumas 2019).

Archaeological sources

Archaeological alchemical remains are extremely rare for the medieval Arabo-Muslim world. They are of two kinds: remains whose context is well known and remains whose context is poorly or not at all known. Unfortunately, the majority of alchemical apparatus that have been preserved are in museums without a specific context, which means that they may have been used for purposes other than alchemy. There might be an exception in Ramla, Israel, where a possible alchemical device was found, but the hypothesis is not very convincing (Gorzalczany and Rosen 2010). Also noteworthy are the finds, associated with a precise context, dating from the late twelfth to the early thirteenth centuries CE in the city of Bilyar in the Middle Volga region (Valiulina 2016). Objects without context are usually very difficult to date and locate, and so must be treated with caution. One example is a glass alembic kept in at the Institut du monde arabe in Paris (object AI 90-08), which is said to date from the tenth or eleventh century in Iran, without any certainty (Institut du monde arabe 1996). Another beautiful glass alembic is kept at the Science Museum in London (object 1978-219), which may have its origin in the Middle East and date from the tenth to the thirteenth centuries. Other glass alchemical instruments are described by Savage-Smith (1997) and Kurzmann (2009), which must always be considered with prudence due to their lack of context. Alchemical devices found in a craft context are also extremely rare for the Arabo-Muslim medieval period. Examples include the aludels in Almadén (Spain), a cinnabar mine where sublimation devices were used to extract mercury from mercury sulfide (HgS), but the site has not yet been excavated, so the information remains doubtful, especially with regard to dating (Hernández Sobrino 1996). Some earthenware vessels, usually found in culinary contexts, may have been used for alchemical operations. For example, pots (*qidr*) from the excavations of the fortress of Shumainis (Syria), dated from the thirteenth and fourteenth centuries, bear specific traces that allow one to postulate that they were used for distillation, but also perhaps for the manufacture of metal oxides (Shaddoud 2017).

More alchemical artifacts are preserved with their contexts from the Latin West before 1500.[7] When the context of alchemical archaeological remains is preserved, we usually encounter three different types: (a) single objects, found alone in a larger context, without a clear connection to alchemical practices; (b) alchemical apparatus related to a craft context; and (c) sets of apparatus not clearly related to a craft context and that could correspond to an alchemical laboratory. Regarding the first category, it is very common throughout Europe to find alembics in domestic contexts without a direct link to alchemy or other crafts (Figure 3.4) (Moorhouse et al. 1972; Thomas 2009; Szymański 2015). Likewise, more and more devices for distillation *per descensum* (see Chapter 2) are being found in seemingly domestic contexts, certainly because they are increasingly recognized (Thomas and Claude 2011). These are pots with perforated bottoms, most often ordinary pots drilled after firing. For distillation, whether *per ascensum* or *per descensum*, the detailed analysis of contexts allows us to glimpse privileged social categories (Figure 3.5). Probably the majority of these discoveries are related to the production of rosewater, alcohol, or various drugs, and these productions seem to be for personal use; in any case, the diffusion of the products is undoubtedly very limited.

The second category consists of alchemical devices, or devices that can be used for alchemy, found in a craft context. This is the case of all the specific devices used to purify or assay metal, such as cupels or crucibles (e.g. in minting or metallurgical workshops; Saussus and Thomas 2019). More interesting are the instruments associated with a potter's workshop, probably for the production of metallic oxides, found in Marseille (thirteenth century; Thiriot 1997; Vallauri and Leenhardt 1997). Other discoveries associated with crucibles could suggest a metallurgical context, such as an aludel and a set of triangular crucibles found in Strasbourg (mid-fifteenth century; Maire and Rieb 1972). The same type of collection – aludels with a large number of crucibles – was found in Beaucaire (Gard, France), in the Saint-Roman-l'Aiguille Abbey (fourteenth century; Leenhardt 1995; Moureau and Thomas 2015), but these two examples might correspond to the third category as well.

The same goes for a large set of apparatus with lute found in Besançon (France), which could evoke an apothecary's workshop, although this attribution is uncertain and the hypothesis of an alchemist's laboratory is also plausible (Goy 1995). Finally, three discoveries are difficult to classify: one from the thirteenth century in Basel (Kamber et al. 1999; Kamber and Kurzmann 2002), another from the fourteenth century in Paris (Rouaze 1986), and a final one from England at Sandal Castle in Wakefield (Moorhouse 1983). In all three cases, the sets show a great variety of objects for distillation

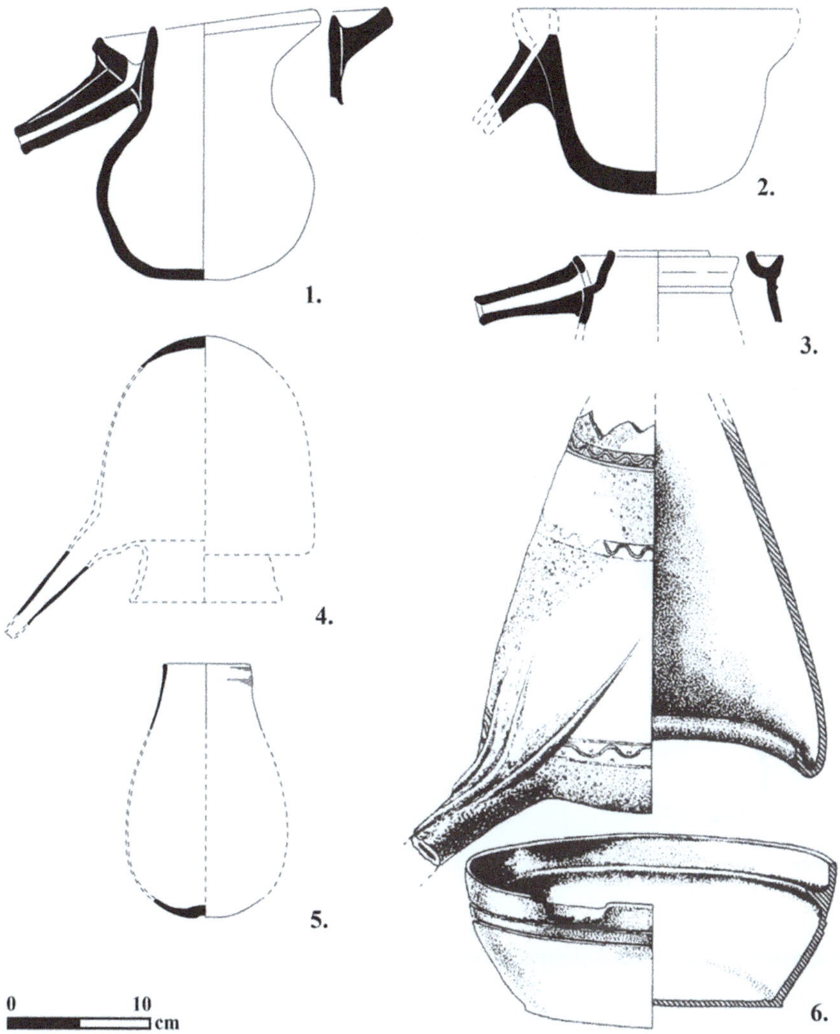

FIGURE 3.4 Archaeological finds of alembics for distillation *per ascensum* (pottery: 1–3 & 6, glass: 4 & 5). (1) Paris (France), Louvre, mid-fourteenth century (drawing by Stephen Moorhouse in Rouaze 1986). (2) Basel (Switzerland), Ringelhof, mid-thirteenth century (Kamber and Kurzmann 2002). (3) Marseille (France), Bourg des Olliers, Sainte-Barbe, thirteenth century (Vallauri and Leenhardt 1997). (4 & 5) Wakefield (UK), Sandal Castle, first half of the fifteenth century (Moorhouse 1983). (6) Köszeg (Hungary), late fifteenth to early sixteenth centuries (Holl 1982).

and sublimation. The crucibles also suggest metalworking, but without a real reference to a craftsman's workshop (Figures 3–4 and 3.6–3.7). We can probably assume that these are laboratories of alchemists: in the artifacts there are devices created specifically for particular operations and devices made to order, but also more common vessels diverted from their primary

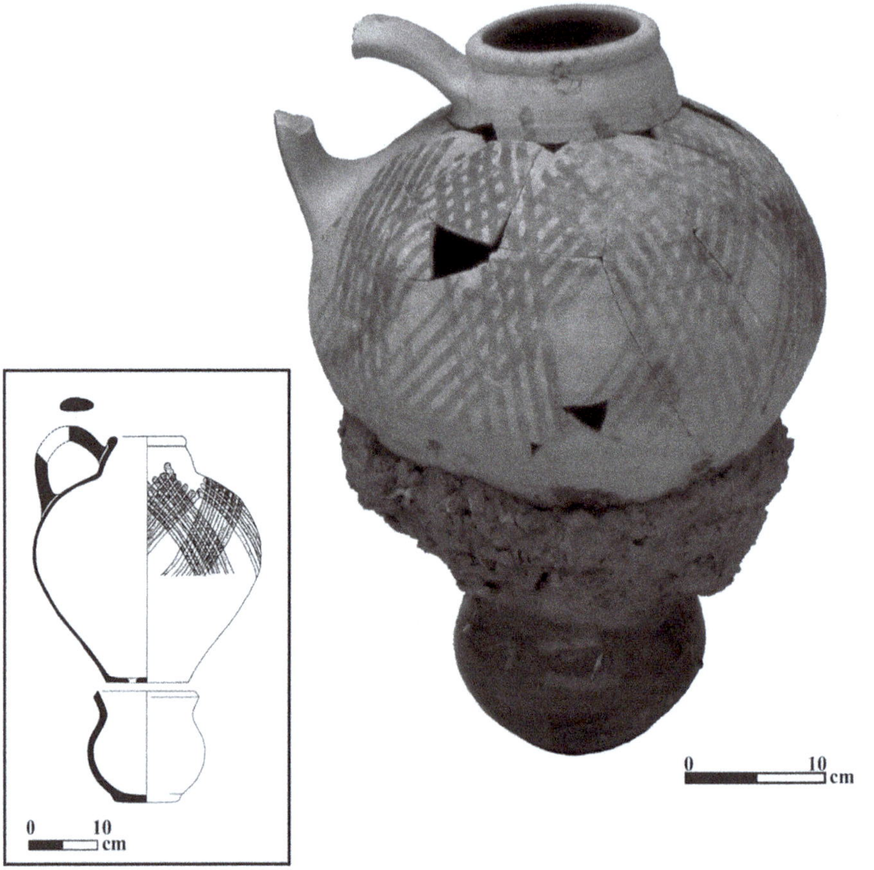

FIGURE 3.5 Archaeological find of two pots luted in a pit for distillation *per descensum* (pottery). Paris (France), Louvre, fifteenth century (drawing and photography by Catherine Monnet in Thomas and Claude 2011).

function with some adaptation. Through these objects, we note the desire to experiment, probably under the influence of alchemical texts, where we find a wide variety of processes. Finally, for other sets, it is even more difficult to draw up hypotheses, as the context or the associated equipment provides a confused picture of the operations carried out (e.g. see a set found at Christ Church, Oxford, in Chadwick et al. 2012).

Robert Anderson has noted that "books and papers have explored telescopes, planetaria, and astrolabes, but little or nothing has appeared concerning furnaces, burning lenses, and alembics" (2000: 5). The recent development of archaeometry since Anderson's remark has made it possible to explore archaeological sources with new tools, especially in the field of metallurgy, to

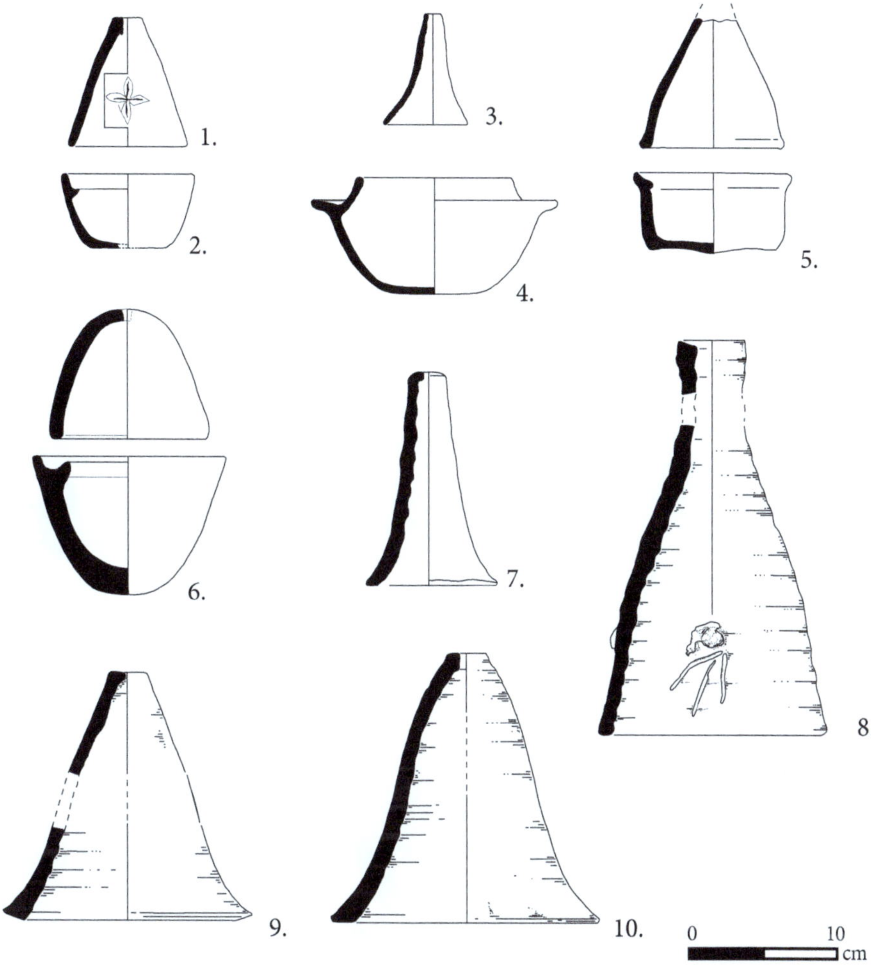

FIGURE 3.6 Archaeological finds of aludels for sublimation (pottery). (1–4) Beaucaire (France), Abbey of Saint-Roman-l'Aiguille, fourteenth century (drawing by Nicolas Thomas). (5) Strasbourg (France), Marais Vert, fourteenth century (drawing by Nicolas Thomas). (6) Basel (Switzerland), Ringelhof, mid-thirteenth century (Kamber and Kurzmann 2002). (7) Paris (France), Louvre, mid-fourteenth century (drawing by Stephen Moorhouse in Rouaze 1986). (8–10) Wakefield (UK), Sandal Castle, first half of the fifteenth century (Moorhouse 1983).

obtain a better knowledge of the processes and equipment used in the alchemical laboratory (Martinón-Torres 2011). However, it must be recognized that it is difficult to specify the objectives and outputs of the laboratories and apparatus discovered. Depending on the context and the assembly of the devices, one is often unsure of their applications to medicine, metallurgy, the manufacture of incendiary compositions, or alchemy, perhaps also because at the time the boundaries between these different fields were not clearly defined.

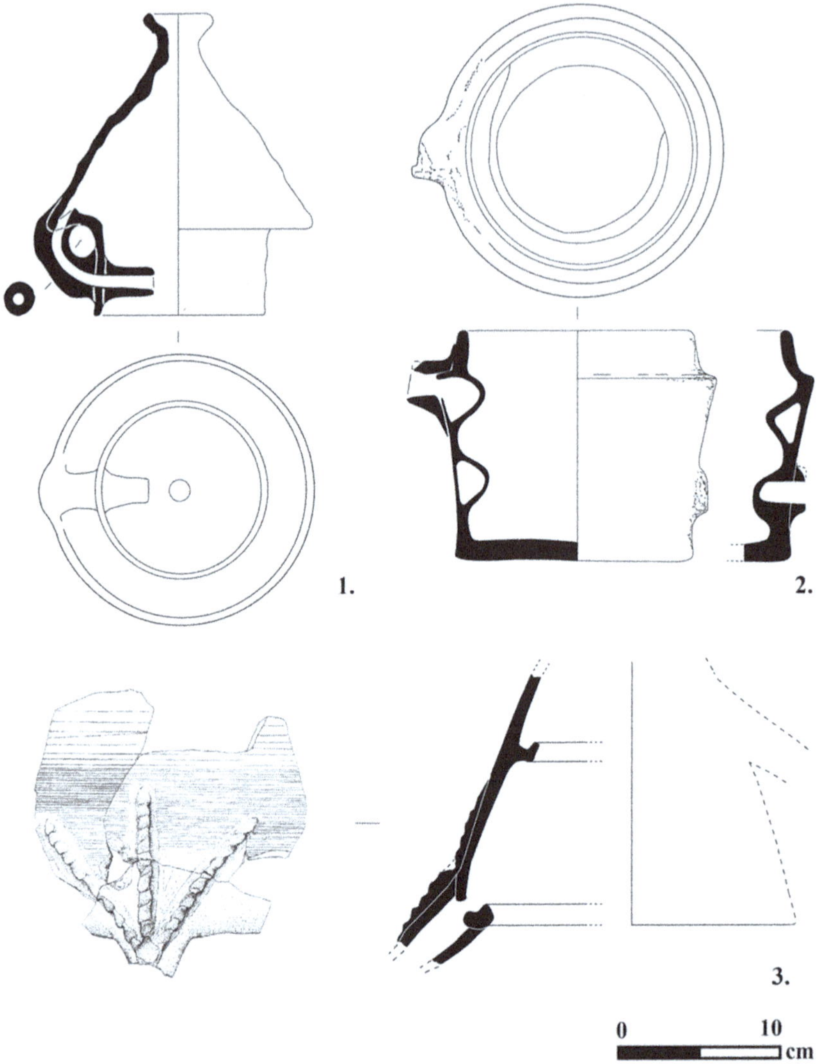

FIGURE 3.7 Archaeological finds (pottery). (1) Pelican, still-head for cohobation or circulation, Paris (France), Louvre, mid-fourteenth century (drawing by Stephen Moorhouse in Rouaze 1986). (2) Worm, apparatus for coiling the still product, Metz (France), Pontiffroy, fourteenth century (Dautremont et al. 2001). (3) Decorated still-head with two collecting channels for fractional distillation, Orschwiller (France), Haut-Kœnigsbourg castle, unpublished (drawing by Thierry Logel, hypothetical reconstruction by Nicolas Thomas).

Experimental sources

Another interesting, yet too rarely used, source of information is experimentation and historical replications (e.g. see Kurzmann 2000; García Moreno and Thomas 2008; Thomas and Claude 2011; Fors et al. 2016; Valiulina 2016). Experimental archaeology often provides important elements for the understanding of medieval instruments in different ways. In particular, this tool makes it possible to produce experimental analogs, manufacturing waste, or apparatus, which can be analyzed and compared with the archaeological artifacts. Experimentation is also a means of verifying hypotheses and validating them, as well as a means of providing a better understanding of the productivity of processes or evaluating the know-how required to manufacture and use these tools.

GENERAL CONSIDERATIONS REGARDING ALCHEMICAL INSTRUMENTS

Alchemical tools can be divided into three main types: tools and accessories, vessels and apparatus, and furnaces. Tools and accessories range from very common instruments, such as hammers, tongs, and surface plates, to more special tools, such as molds, filters, and lutes. They are made from a variety of materials, depending on the tool, from earth and metal to linen (for filters). Vessels and apparatus are the containers used by alchemists, from basic pots and crucibles to more sophisticated alembics and aludels. They were most commonly made of fired clay or earth with additional ingredients (grog or chamotte, vegetable fiber, etc.). Glass was often described by alchemists as the best material for specific vessels such as alembics or aludels; it won favor because of its transparency and supposed purity (Barthélémy 1995; Thomas and Rodrigues 2017). However, because of its price and fragility, especially at high temperatures, fired clay was a much more common material. Metal was sometimes used for instruments. Although it was seldom mentioned in technical treatises, it often appeared in inventories, especially for alembics whose heads were often made of lead (Bénézet 1999). Sometimes metal was avoided because of its interaction with the ingredients and products of the processes, particularly mercury and its compounds. The term "furnace" applies to any device or place where alchemists heated their vessels with fire. They were made of fired earth or clay, bricks, or stones.

The question of how alchemical instruments were made can be asked at two levels: from a historical and general point of view, namely how these devices were discovered and developed, and from an individual and material point of view, namely how these instruments were made in the daily life of an alchemist.

From the historical perspective, it is complicated to date precisely the invention of devices as old as alembics and aludels (not to mention basic tools such as hammers, etc.). Metallurgy dates back to the fourth millennium BCE, and the oldest known apparatus that could have been used for distillation or sublimation and that resembles alembics and aludels was found in Tepe Gawra in Mesopotamia and possibly dates back to 3500 BCE (Levey 1955). Arabo-Muslim alchemists and later Latin alchemists mostly inherited their equipment from earlier periods and, although they improved them and above all adapted them to their specific processes, the general tendency was toward conservatism, as can be seen in crafts in general, where a high degree of stability is a common feature of the processes and tools.

Regarding the manufacture of instruments, we have only a few indications, even when comparing textual, archaeological, iconographic, and experimental sources. Nevertheless, three situations can be distinguished. In some cases, the user of the instrument was the maker. Indeed, we find in some texts detailed recipes on tool-making, such as the description of the fabrication of an aludel in Rāzī's *Kitāb al-asrār* (Dānish-Pazhūh 1964: 11; English translation in Moureau forthcoming) or in the *'Ayn al-Ṣan'a* (Stapleton and Azo 1905: 62–4). This is especially true for furnaces, as their making required only basic skills. The second possibility for the alchemist was to obtain instruments made by craftsmen. This is the case for glass equipment: glassmaking was a difficult art that required great know-how and long experience. Metalworking also presupposed acquired abilities, and the making of metal instruments, such as alembics, was therefore a matter for craftsmen. In the case of fireclay, there is archaeological evidence of manufacturing failures that show that potters made crucibles or alembics (e.g., see Miller and Stephenson 1999: 25–6; Guadagnin 2007: 78, 111, 192, 548–52). The last possibility is the customization of ready-made instruments. For instance, in the abovementioned Rāzī's description of how to make an aludel, if the head of the aludel is indeed entirely made by the alchemist, the lower part is a common cooking pot (*burma*) sawn into two pieces (Moureau and Thomas 2015). Many items of alchemical apparatus were actually borrowed from the kitchen or other crafts and then adapted to alchemical processes. For the distillation *per descensum*, archaeological findings show that common ceramics were adapted after firing by perforation and thus reused for this specific purpose (Thomas and Claude 2011).

Archaeological findings of devices give an idea of the dimensions of these objects, which are generally very small. For example, in the case of distillation *per ascensum*, it can be observed that the vessels can only contain a small amount of liquid to be distilled, from less than a liter to a few liters at the most. For distillation *per descensum*, the amounts are sometimes larger, up to about 10 l, but in all cases the quantities of distillate produced are very small. To increase production, the number of vases in the same furnace is multiplied, as can be seen in a representation in MS Paris, Bibliothèque nationale de France, arabe 2187, fols 103–4 (fifteenth century) of an abovementioned treatise of Shams al-dīn al-Dimashqī (Institut du monde arabe 1996: 196) or in late medieval treatises such as that of Brunschwig (1500).

PLACES OF WORK: LABORATORIES AND WORKSHOPS

It is difficult to talk about the alchemical laboratory in the Middle Ages without running the risk of being anachronistic. The modern definition of such a place probably cannot be applied to this period. Moreover, the Latin term *laboratorium* does not usually designate the place where the alchemist worked. On the other hand, and despite the recommendations of Pseudo-Albert in the *De alchimia* (i.e. "an alchemist must live, far from men, in a private house, in which there are two or three rooms exclusively intended for sublimations, solutions and distillations"; Borgnet and Borgnet 1908: 549), the clear separation between the place where one lives and the place where one works can also be seen as an illusion. As regards the laboratory – the place where chemical experiments were conducted – a clear differentiation appeared very late, in the seventeenth or even eighteenth century only (Shapin 1988). This may also apply to the workshops of craftsmen, for whom workshops and houses are often merged. Finally, we should also beware of the romantic vision of the alchemical laboratory popularized by art, especially the Flemish painters of the sixteenth and seventeenth centuries, whose most famous representations of the alchemist at work are those of Pieter Brueghel the Elder in the sixteenth century or David Teniers the Younger a century later (Hill 1975). The latter's paintings often show an elderly man, sometimes alone, sometimes with a few helpers, in a dark place where there is disorder and dirt (Holmyard 1956). In the same way, one should not consider as realistic the representations showing the alchemist distilling in a garden, as in the *Liber de arte distillandi de simplicibus* by Hieronymus Brunschwig (1500). The alembic in the middle, of a controlled and rationalized nature, is only there to serve the purpose of this surgeon and apothecary who wished to sell his distilled waters in Strasbourg to the greatest number of people. It is much more likely that Brunschwig's furnaces were in the back of his shop near the Strasbourg fish market (Bachoffner 1993).

This being said, for the periods after the Middle Ages, several researches have focused on the place where people experimented, its architecture, the organization of its space and work with the practice, its diffusion, the modes of learning, the culture of secrecy, and more generally on the social conditions of work (Beretta 2002: 286–311; Nummedal 2007: 119–46; Kohler 2008).

Sources are lacking to characterize the workspaces of the alchemists during the Middle Ages. While the sources cannot precisely define these workspaces and their organization, they can still provide some valuable insights. First, there are archaeological sources that help to clarify the context. For example, abbeys often supplied distillation apparatus, especially in England (Moorhouse et al. 1972; Tyson 1996; Willmott 2002). One can imagine that these apparatus were in use in common buildings and probably in apothecaries. The abovementioned set of aludels and crucibles found in a context related to the abbey of Saint-Roman near Beaucaire (Leenhardt 1995) indicates metalworking, possibly related to mercury and cinnabar, which might suggest a link with the manufacture of pigments for the scriptorium.

Other findings come from more aristocratic contexts, in castles. For example, in England at Sandal Castle, all the artifacts reveal the practice of distillation and sublimation associated with small metallurgical crucibles during the first half of the fifteenth century (Moorhouse 1983).

However, many of the alchemical devices come from urban contexts. For example, the set of vases dating from the second half of the fourteenth century found in Paris, in the Cour Napoléon of the Louvre, testifies to the presence of an alchemist in the gardens of the Saint-Nicolas school (Rouaze 1986). The set included glass and ceramic alembics, aludels, and crucibles for small metalwork. One of these finds was a "pelican," the only apparatus of the period used for cohobation (see Chapter 2; see Figure 3.7). If this activity cannot be related to the teaching given in such an establishment, it was likely practiced by a canon. In any case, this discovery attests to a literate, possibly scholarly, and perhaps even religious milieu. For the set of apparatus found in Basel, it is perhaps to be associated with a rich noble family of the thirteenth century, the von Butenheims (Kamber and Kurzmann 2002). Most alembics for distillation *per ascensum* found in Europe and dated to the Late Middle Ages come from urban contexts – usually of the more wealthy echelons of society. Alembics found alone are likely to have been related to rosewater or alcohol production in a domestic context (Thomas 2009).

For certain operations, such as distillation *per descensum*, the association of devices with urinals would indicate a use in a medical context (Thomas and Claude 2011). Here again, urban contexts indicate literate environments, but these instruments seem to have been used only for occasional distillations, as were those of distillation *per ascensum*, and in any case they were far less used than those of an apothecary. These findings suggest the manufacture of medicines for personal use. It therefore makes little sense to research, or even to imagine, a practice in a "laboratory" as we understand it today.

PRACTICES, TOOLS, AND APPARATUS

In this section, some important alchemical instruments and tools are described in detail. At the end of this section, the reader will find a list of instruments that can be found in an alchemical laboratory, accompanied by the words that designate them in Arabic and Latin.

Lute

Lute is a material, hence is not strictly speaking an apparatus, but it is indispensable for the use of alchemical equipment. It is a coating medium used for covering pots to protect them from fire, to seal vessels, and to join parts of apparatus (Thomas 2013). In the Middle Ages, it was usually made of earth, clay, plaster, or sometimes flour, or else mixed with organic temper (straw, dung, animal hair, etc.) and moistened with a liquid (water, egg white, etc.). Some lutes were refractory (resistant to heat), and recipes varied according to the purpose of the lute. This very common material was strangely highlighted in alchemical texts; alchemists provide numerous lute recipes, which leave the modern reader perplexed. The lute was called in Arabic "lute of wisdom" (*ṭīn al-ḥikma*), translated into Latin as *lutum sapientiae.*

Alembic

The word "alembic" comes from the Latin *alembicum*, which is a transcription of the Arabic word *al-anbīq* or *al-ambīq*, which itself comes from the Greek term ἄμβιξ (*ambix*). The device has already been partially described in Chapter 2, along with the process of distillation *per ascensum*, for which the alembic is designed. Medieval alembics were usually made of fired clay or glass, but also of copper or lead. At first, the word alembic designated the head of the device, but it often came to refer to the entire apparatus. It consisted of a cucurbit (*qar'/cucurbita*; i.e. a gourd-shaped pot), on top of which was placed a head (or helmet), an inverted cup, usually with a collecting channel in its bottom, and most often with a spout or pipe. At the end of the pipe was placed a receiver (*qābila/receptaculum*). The material to be distilled was placed in the cucurbit, and all the joints were luted. The device was gently heated in a special medium – a bath with ashes, sand, water or something else – in order to avoid too high a temperature or thermal shocks that would damage the distillation. The alembic without a pipe was called a "blind alembic" (*anbīq a'mā/alembicum caecum*).

In the Middle Ages, in the West, at least two major innovations in the art of distilling should be mentioned. Concerning the cooling system, the introduction of the coil or worm – a pipe passing through a container of cold water – appears to date at least from the thirteenth century. It is mentioned by Theodoric of Cervia (1205–1298) when he distinguished the distillation *serpentis* from the distillation *canonis* (Halleux 1981: 249), and also by the physician Taddeo Alderotti (1223–1303) in the *Concilia medicinalia* (Crombie 1953: 103). It is represented in a manuscript of the fifteenth century (MS London, British Library, Sloane 1698, reproduction in Lennep 1984: 11). Some have seen in Zosimos' texts another cooling system for the alembic head, of which the Moorish-headed alembic is said to be a representative in the Arabo-Muslim world (Taylor 1945; Mertens 1995: cxxix). However, the practice of cooling the head with a wet cloth continued for a very long time, even until today (Thomas 2009). An archaeological discovery of a coil in a potter's workshop dated between the middle of the thirteenth century and the end of the fourteenth century shows a terra-cotta vessel with a pipe along the wall from top to bottom (see Figure 3.7).

The other innovation is fractional distillation. It consists in collecting the vapors that condense at different heights in the alembic. In this way, the "heavy" and "subtle" vapors can be separated. A representation of such an apparatus is given in the *Ordinal of alchemy* (MS London, British Library, Add. 10302; see Figure 3.3). A fragment of an alembic head from the late fifteenth or early sixteenth century found during former excavations at the Haut-Kœnigsbourg castle in Alsace (France) may have been designed for fractional distillation (see Figure 3.7).

Aludel

The aludel is the apparatus dedicated to sublimation (Moureau and Thomas 2015). It has already been partly described in Chapter 2. The word "aludel" is the Latin transcription of the Arabic term *al-uthāl*, which itself comes from the Greek αἰθάλη/αἴθαλος (thick smoke, soot) or αἰθάλερον (aludel). There were three main types of aludel in the Middle Ages. The simplest aludel was a bottle with a long neck. The material to be sublimated was placed in the bottle and the opening of the bottle was sealed with lute; a small hole was drilled in this lute plug to allow the build-up of pressure in the bottle to escape. The sublimate settled on the inner side of the neck. This system was very simple; it was known from early times in the Arabo-Muslim world and was already in use in the Latin West before the penetration of Arabic knowledge, since it was already described in the *Mappae clavicula* in the ninth century (Smith and Hawthorne 1974). A second type of aludel is only found in Arabic texts and never appears in Latin sources. It is described in detail in Rāzī's texts and in the *'Ayn al-ṣan'a* (Rāzī, *al-Madhkhal al-ta'līmī*, in MS Rampur, Raza Library, kīmiyā" 12, fols 95v–6r, English translation with inconsistencies in Stapleton et al. 1927: 356–7; Rāzī, *Kitāb al-asrār*, in Dānish-Pazhūh 1964: 11, English translation in Moureau forthcoming; *'Ayn al-ṣan'a*, in Stapleton and Azo 1905: 62–4, 69–70). It was a complex device made of a sawn-off piece of a cooking pot (*burma*). On the top of this pot was attached and luted a clay disk in which a gutter had been carved. A lid called a *mikabba* (literally something "inverted"), similar to a tajine lid, was placed on the disk. Two small "wings" (i.e. small pieces of lute) were attached on the sides of the pot in order to suspend it in the furnace. The material to be sublimated was placed in the cooking pot and sublimed on the inner side of the lid and in the gutter of the clay disk. The third type was also an aludel consisting of a pot and a lid, but without the clay disk; it is described in detail in Pseudo-Geber's *Summa perfectionis* (Newman 1991: 377–85, 687–92). The upper edge of the lower pot was shaped like a collecting channel into which the lid of the upper part was inserted. The joint was not luted, which allowed the lid to be changed during the process when enough sublimate had collected on the inner side of the lid.

Descensory

The process of descension (see Chapter 2) was conducted in a device called a descensory: the Latin terms are *descensorium*, *chimina*, from the Arabic *qinnīna*, "bottle," and *botus barbatus*, a transcription from the Arabic *būṭ bar būṭ* or *būt bar būt*, itself a borrowing from the Persian *būt bar būt*, meaning "crucible on crucible." The usual descensory consisted of two pots, one placed on the top of the other; in the *Summa perfectionis*, the descensory refers only to the upper pot, and this pot had to have a narrow bottom (as a reversed bottle, hence the word *chimina*) and be very smooth inside in order to allow the metal to flow easily. Holes (or a single hole) were drilled in the bottom (preferably narrow) of the upper pot. The joint was luted, and a lid could be placed above the upper pot. The descensory used in the descension was very close to the apparatus used in distillation *per descensum* (see Chapter 2); the author of the *Summa perfectionis* even stressed the link by advising to use the *chimina* (the descensory) in distillation *per descensum* (Pseudo-Geber, *Summa perfectionis*, *cap.* 46; Newman 1991: 408). However, the processes were very different, since descension required a high temperature whereas distillation required a low temperature. Archaeological evidence shows us that the devices, although they looked very similar, were actually different: the descensory consisted of crucibles, while the apparatus for distillation *per descensum* was usually made up of simple pots, often common cooking pots (see Figure 3.4).

Furnaces

A furnace was any device used to heat a material, usually by fire (sometimes by fermenting dung). It ranged from a simple pit dug in the earth to much more complex structures. They were usually made of clay or earth, or brick or stone. Most furnaces for alchemical operations were natural draught furnaces, but bellows (*minfakh*, *ziqq*, *kīr*/*follis*, *barchinus*) could be added when a higher temperature was needed (for melting metals with higher melting points, for instance). The fuel could be wood, or animal dung when wood was scarce, or charcoal, which allowed a higher temperature to be reached.

An impressive variety of furnaces are described in alchemical sources. The Arabic language itself reflects this diversity, as we can list no less than seven names for furnaces: *kūr*, *tannūr*, *atūn*, *mustawqad* or *mawqid*, *ṭābadshān*, *kānūn*, and *nāfikh nafsi-hi*. If some of these words, such as *kūr*, *tannūr*, or *atūn*, were quite generic terms, others referred to very specific furnaces, such as the *mustawqad*, a cylindrical stove used for sublimation. The furnace for cupellation in Arabic literature, which pertained to craftwork more than to alchemy, was also considered an unusual furnace, and the bellows used for oxidizing the lead were called *rawbāsh* or *rawbāṣ*, which blew down directly on the cupel, as in a forge (Moureau and Thomas 2016: 104). The same variety of furnaces appears in Latin literature, where many furnace descriptions are found, although there is not the same multiplicity of terms for this object in the Latin language.

Tools and Accessories

File (*mibrad*/*lima*)
Filter (*rāwūq*/*liquatorium*, *feltrum*)
Funnel (*qimʿ*/*infundibulum*)
Hammer (*miṭraqa*/*malleus*)
Ladle (*mighrafa*, *milʿaqa*, *māshū*/*coclear*)
Linen cloth (*khaysh*/*pannus lineus*)
Lute (*ṭīn*/*lutum*)
Marble (*ṣalāya*/*marmor*)
Mortar (*hāwan*, *mihrās*/*mortarium*)
Mold (*misbaka*, *rāṭ*/*modulus*, *forma*)
Pestle (*nisāb*, *midaqq*, *mukassir*/–)
Round mold (*kura*/–)
Shears (*muqaṭṭiʿ*, *miqṭaʿ*/*forfex*)
Sieve (*munkhal*/*cribrum*, *taratantara*)
Stone used to grind on the marble, marver (*fihr*/–)
Tongs (*māsik*, *kalbatān*, *ambur*/*forceps*)

Vessels and Apparatus

Alembic (*ambīq*, *anbīq*/*alembicum*)
Aludel (*uthāl*/*aludel*, *alutel*) with its special lid (*mikabba*/*coopertorium*)
Basket (*salla*/various words) or cage (*qalaṣ*/–; for inhumation)
Blind alembic (*anbīq aʿmā*/*alembicum caecum*)
Bottle (*qinnīna*/*ampulla*, *vas*)
Cauldron (*mirjal*, *ṭinjīr*/*caldarium*)
Clay box in which materials are calcinated (*durj*/–)
Cooking pot (*burma*/*olla*, *vas*)
Crucible (*būṭaqa*, *būṭ*/*botus*, *crucibulum*)
Cucurbit (*qarʿ*/*cucurbita*)
Cup, bowl, plate (*qadaḥ*, *kūz*, *jām*, *bāṭiya*, *sukurruja*, rarely *riṭl*/*vas*, *cauchia*, *scutella*, *phiala*)
Cupel (*kūkh*, *kūḥ*, *kūj*/*cupella*)
Descensory (for distillation *per descensum*) (*mustanzil*?, *būṭ bar būṭ*?/*descensorium*)

Descensory, "crucible on crucible" (*būṭ bar būṭ/botus barbatus, descensorium, chimina*)
Frying pan (*miqlāt/paella, sartago*)
Jar with lid (*barniyya/olla, vas*)
Phial (*qārūra/ampulla, phiala*)
Pot (*qidr/olla, vas*)
Receptacle (*qābila/receptaculum*)

Furnaces

Bellows (*minfakh, ziqq, kīr/follis, barchinus*)
Bellows (for cupellation; *rawbāsh, rawbāṣ/–*)
Furnace (*kūr, tannūr, atūn, mustawqad, mawqid/furnus, fornax, furnellus, athanor*)
Furnace heated from above (*ṭābadshān, kānūn/–*)
Furnace with perforated sides = induced draft furnace (*nāfikh nafsi-hi/–*)

CONCLUSION

For the medieval period, it may be difficult to efface the image of the alchemist in his laboratory popularized much later in paintings and up to the present in literature or cinema. The risk of anachronism is very real, since the intentions behind these images are not so much to depict a reality as to pass judgment on alchemy, most often in favor of a "modern chemistry" in search of a completely new rationality. In fact, the data for reconstituting this laboratory are very scarce, which is why it is necessary to mobilize various sources in order to go beyond the commonly accepted platitudes and to approach, perhaps, a practice somewhat misleadingly stamped with the seal of secrecy.

The medieval alchemical laboratory is mainly known through tools and apparatus. From this point of view, Latin alchemy offers more data than the Arabo-Muslim world. It should be pointed out that, unlike processes, alchemical instruments can be more easily understood by the iconography and archaeological sources than texts. Archaeological discoveries make it possible to put textual descriptions and images of the apparatus into perspective. However, although archaeological sources allow a more reliable approach to the apparatus, they often remain subject to doubt, as it is rare to be able to distinguish between an instrument of an alchemist and an instrument of a craftsman, for the reason that these tools are often identical. An accurate and complete analysis of the discovery context is necessary to resolve these uncertainties, but unfortunately the data do not always allow for this. On these issues, research increasingly oriented toward the analysis of workshop and laboratory materials and residues should make it possible to resolve certain questions in the future. The development of archaeometry, particularly in the field of organic chemistry, is very promising.

Nevertheless, some general conclusions can be drawn from the available data. Like practices and experiments, the apparatus of medieval alchemists are characterized above all by stability. The Middle Ages were more conducive to adaptation and improvement than to innovation, without forgetting, however, that some radical enhancements were introduced during this period, especially for distillation. Medieval devices, whether alchemists' or craftmen's, were always small. The quantities produced were modest, and mass production had not yet made its appearance. Another particularity of the equipment, this time only alchemical, is its great adaptability. While there have obviously been instruments directly produced for alchemical operations, these apparatus are marginal and not common. They were produced to order or, when possible, manufactured by the user himself. The forms appear to be constructed from models derived from the texts in circulation, and probably from the many images found in manuscripts. This particular mode of transmission is certainly at the origin of the great variety of shapes and types of objects, even if they remain similar in the principles of their action. Adaptability can also be seen in the reuse or modification of common vessels, such as simple pots from the kitchen, which are then transformed into alchemical devices.

Adaptability, mode of transmission mainly through texts rather than objects or orally, marginal practice, and very limited production capacities – these are characteristics that eventually separate alchemy from craft, laboratory from workshop, and knowledge from know-how.

CHAPTER FOUR

Culture and Science: *Alchemy's Scientific Contexts and Critiques*

REGULA FORSTER AND JEAN-MARC MANDOSIO WITH
ANTOINE CALVET AND GABRIELE FERRARIO

ISLAMICATE WORLD[1]

Regula Forster

Alchemy must have been a very widespread practice in the Islamicate world, given the vast number of existing manuscripts and the even larger number of texts known to have existed (and not yet tracked down in the libraries of the world). However, alchemy was not a practice that existed in a void, since it had close relations to other fields of science and scholarship.

Categorization of the sciences

Authors writing in Arabic were very interested in categorizing the sciences (Hein 1985). Perhaps the usual way of looking at sciences and knowledge was to ask about the origin of the science in question and to categorize them as being either of Arab or of foreign origin (*'ulūm 'arabiyya* vs *'ulūm al-'ajam*).[2] In this approach, Abū 'Abd Allāh Muḥammad b. Aḥmad al-Kātib al-Khwārizmī (fl. late tenth century), in his *Mafātiḥ al-'ulūm* ("Keys of the sciences"), dedicated to a Samanid vizier, discussed alchemy as one of the foreign sciences, alongside philosophy, logic, medicine, arithmetic, geometry, astronomy, astrology, music, and mechanics (Sabra 1976).

In his tenth-century bookseller's catalog, *al-Fihrist* ("The catalog," ed. Tajaddud 1393 AH/1973 CE; trans. Dodge 1970), Ibn al-Nadīm treated alchemy in a separate treatise: though he announced that he would put the remaining arts here, into the last tenth section, it contains writers and works only on alchemy. This shows both the problem of categorizing this art and its importance, as it is granted a separate treatise while, for example, medicine is part of the section on philosophy, mathematics, and astronomy. It is also noteworthy that alchemy was not put into the eighth section, together with magic.

Other scholars had problems with the treatment of alchemy, as this art had not yet been invented in Aristotle's time and therefore he had not discussed it. The issue was handled differently by scholars around the mid-tenth century. The mid-tenth-century philosopher al-Fārābī did not mention alchemy in his *Iḥṣā' al-'ulūm* ("The enumeration of the sciences"), thereby ignoring the problem rather than solving it. This is rather surprising, as he does enumerate practical sciences such as engineering (Rudolph 2012b: 378–9). Interestingly, he does acknowledge alchemy's existence and even calls it a necessarily existing science in his *Maqāla fī wujūb ṣinā'at al-kīmiyā'* ("Treatise on the necessity of the art of alchemy," ed. Sayili 1951; Ullmann 1972: 250; Rudolph 2012b: 393). Ibn Farīghūn (fl. mid-tenth century), in his compendium on the classification of the sciences, *Jawāmi' al-'ulūm* ("The comprehensive of the sciences"), mentions alchemy among those sciences whose truth is disputed, alongside the science of talismans, physiognomy, astrology etc. That he sees alchemical practice as being feasible, at least in theory, becomes clear in his explanation that "alchemy does what Nature does, which is the act of the Creator" (Ibn Farīghūn 1985: 146).

The *Rasā'il Ikhwān al-ṣafā'* ("Epistles of the Brethren of Purity," written probably in the mid-tenth century), a famous encyclopedia, lists alchemy among the *'ulūm riyāḍiyya* ("mathematical or propaedeutic sciences"), at the same level as reading, writing, grammar, lexicography, calculation, poetry, magic, etc. (seventh epistle, Ikhwān al-Safā' n.d.: I: 266–7). Their common aim is that, through these sciences, one can gain a living and eventually achieve a good life, as opposed to the conventional sciences of the *sharī'a* (i.e. religious studies) and the true philosophical sciences, which do not serve a practical end in this world. In Epistle 52 (Ikhwān al-Safā' n.d.: IV: 286–7 and 305; trans. Callataÿ and Halflants 2011: 95–6, 143), alchemy was counted as part of magic (*siḥr*), alongside astrology, talisman-making, medicine, and, in the first passage, asceticism. This, therefore, is a completely different categorization compared with that of letter 7, which stresses the occult aspects of the divine art (i.e. alchemy; Callataÿ and Halflants 2011: 10–14).[3]

In the eleventh century, Ibn Sīnā (Avicenna) divides natural philosophy into two classes: *uṣūl* ("fundamentals") and *furū'* ("branches"). This distinction shows the importance of Aristotle for Ibn Sīnā, as *uṣūl* are those branches of knowledge already treated by Aristotle, while *furū'* are those sciences not present in the Corpus Aristotelicum, most notably medicine, astrology, and alchemy (Ibn Sīnā n.d.: esp. 111). Similar classifications were offered by Ibn al-Akfānī in his fourteenth-century work *Irshād al-qāṣid* ("The guidance for the seeker") and, in the sixteenth century, by Tāshköprüzāde in his *Miftāḥ al-sa'āda* ("The key to happiness"; ed. Witkam 1989: 45–53; Ṭāshköprüzāde n.d.: I: 301–22; see also Artun 2013: 21–3).

Alchemy and other occult sciences

Alchemy has a strong relationship with other natural and especially occult sciences. Closest perhaps are the links with astrology and magic, whose methods are sometimes used in alchemy (Ullmann 1972: 145; Carusi 2000), as is the case in *Kitāb al-aṣnām al-sab'a* ("The book of the seven idols"), attributed to Apollonius of Tyana (Karimi Zanjani Asl 2013). As Lippmann has argued, it is not surprising that star constellations were sometimes seen as fundamental for a successful completion of the alchemical work, as they were also seen as central for many other actions, like marriage, conception, etc. (1919: I:322) Already in the first half of the ninth century, Sahl b. Bishr has argued that the star constellations should be taken into account before practicing alchemy (Burnett 1992: 105–6). The same line seems to be present also in Ottoman alchemy (see Artun 2013: 164). Not surprisingly, the Jābirian corpus includes texts on astrology, most notably the *Kitāb al-baḥth* ("Book of research"; see Kraus 1942–1943: I:142–6, no. 1800). In the thirteenth century, Jābir is mentioned as an astrologer by 'Alī b. Ṭāwūs al-Ḥillī (d. 1266; Haq 1994: 19).

Connections with magic are also often to be observed in alchemical texts (see Anawati 1996: 880–2). There is, for example, a shared idea of the importance of names and letters and that to change the order of letters of a certain substance would lead to a transmutation (see Carusi 2000: 465). In Maslama al-Qurṭubī's *Ghāyat al-ḥakīm* and *Rutbat al-ḥakīm*, al-Qurṭubī places alchemy below magic, explaining that to learn about alchemy would be the last step toward achieving the secrets of magic, which in itself were the goal of the sage (Callataÿ and Moureau 2016).

Interestingly, the use of a so-called elixir or philosophers' stone as a panacea is not only attested in alchemical writings, but also in a genre that has been called treasure hunter manuals (Braun 2017). In one of these works, extant in a manuscript of the fifteenth century, the treasure hunter is advised to go to a certain church in Cairo. There he will eventually find a zizyphus tree with its fruit:

> If you open one of these fruit into halves, you will find one half red and the other one yellow. They free from leprosy, elephantiasis and blindness. If you bring the weight of one part (*juz'*) of it unto a thousand *mithqāl* of mercury, it becomes an elixir (*iksīr*). And if you rub it in oil and if a sick man oils himself with it, he will be cured. If a blind person uses it as a collyrium, God will let him see the light again.
>
> (Kamāl 1907: I: 2, trans. RF)

Several points are striking: first of all, the basis of the procedure is not a mineral, but a vegetable, namely the fruit of the zizyphus tree, a tree famous for growing in paradise (cf. Q 53:14; for interpretations thereof, see Forster 2001). Second, this has to be combined with mercury, a substance that is fundamental to alchemy, being at the core of the mercury/sulfur theory.[4] Finally, we might detect here some calque of the ancient usage of the word *iksīr*, "elixir," which seems to have been, before becoming a synonym of the philosophers' stone, used for a collyrium – that is, a drug used in the treatment of the eyes (Strohmaier 2016: 425). In this citation, the elixir is obtained not in an alchemical but in a more or less pharmaceutical way, by mixing the fruit with mercury or with oil. While its effectiveness as a collyrium is explicitly stated, it remains unclear for what kind of illnesses the oil infusion and the mercury compound can be used – quite likely on all kinds of diseases. If that assumption is correct, this citation therefore would refer to a kind of magically obtained elixir-cum-panacea.

Alchemy and medicine

A close relationship of alchemy and medicine has already been noted by Lippmann, who remarked that the elixir was seen as a panacea already in ancient Egypt (Lippmann 1919: 65). However, he also noted that in a Syriac work published by Berthelot, no connection between alchemy and medicine is made, but pharmacology is treated in the framework of citations taken from Dioscorides and Galen (Lippmann 1919: 92). Manfred Ullmann went so far as to state that the relationship between alchemy and medicine was only an external one, consisting mainly in the influence of Galen's writings upon alchemical thought, while even in the production of mineral drugs, alchemical methods were of no importance (1972: 150, n. 3). Indeed, the link between alchemy and medicine became much more prominent only in early modern times and under the influence of Paracelsian iatrochemistry, which meant that drugs were now produced by alchemical processes, mainly through distillation (Bachour 2012).

However, links between medicine and alchemy can still be detected. Generally speaking, alchemy seems to share some of two important features of medicine: not having been treated by Aristotle and being potentially useful to whoever masters it. Just like medicine, alchemy is more similar to an art rather than a science. Paola Carusi (2003) has emphasized the following points linking alchemy to medicine: texts on alchemy are likely to cite the same authorities as texts on medicine; for instance, Hermes and Asclepius might appear in both kinds of writings. Alchemical and medical texts are similarly structured: the alchemical work is seen as a kind of healing of the metals, described in the same terms as the cure of a patient. The goal of both the alchemist and the physician is the amelioration of the state of the treated, be it the patient or the metal. In their allegories, alchemical authors make frequent use of medical images: they speak about the development of the embryo (meaning the ripening of the metals), of the birth (meaning the creation of gold), etc. For both alchemy and medicine the primary qualities (hot, cold, dry, and moist) are central; their relations to each other are a concern of both. The alchemist, like the physician, cannot do anything against Nature, but can only assist Nature.

The similarity of alchemy and medicine is underlined in several works of the prominent fourteenth-century alchemist al-Jildakī.[5] He goes so far as to explain that medicine is a specific part of the world of alchemy:

> Medicine, however, is a branch of the alchemical world, but a specific one, because the science of medicine is concerned with the treatment of the bodies (*abdān*) of the animals. ... Likewise, in the alchemical world, the treatment of the bodies (*ajsām*) [is at stake].
>
> (al-Jildakī, *Miṣbāḥ*, MS Leiden, University Library, Or. 1274, fol. 13r)

Medicine will heal the animal bodies (*abdān*), while alchemy is concerned with the treatment of the inorganic bodies (*ajsām*) of the metals, according to al-Jildakī. Generally, he often uses medical metonymies in explaining alchemical procedures and functions. For example, he calls the fusion of spirit (*rūḥ*) and soul (*nafs*), most likely the fusion of mercury and sulfur, the "liberation from illness" (*al-khalāṣ min al-dā'*), the process of *tazwīj* ("marriage," a kind of combination or fusion; see Ullmann 1972: 264–5) is called the "source of healing" (*'ayn shifā'*), and, as a citation from Jābir, the elixir "the physician of the sea" (*ṭabīb al-baḥr*; al-Jildakī, *Kitāb ghāyat al-surūr fī sharḥ al-shudhūr*, MS Leipzig, University Library, 836, fols 7r, 7v, 14v).

Given these important conceptual parallels between alchemy and medicine, it cannot come as a surprise that the elixir was at times seen as a panacea. This idea seems to have played a minor role in Greek alchemy and it is not very important in Syriac works. It might have been more prominent in Egypt, but that is difficult to evaluate. However, it was quite prominent in ancient China, and therefore might have been reintroduced into the Middle East of Islamic times through contacts with China after the extension of the Islamic Empire toward the east (Strohmaier 2016: 430).

While the transmutation of metals is at the core of Islamicate alchemy, as noted by Manfred Ullmann (1972: 257) more than forty years ago, the use of the philosophers' stone as a panacea is far less prominent. However, it is mentioned already in the corpus of writings attributed to Jābir b. Ḥayyān, one of the alleged founding fathers of Arabic alchemy. In the *Kitāb al-khawāṣṣ al-kabīr* ("The great book on the specific qualities"), the author boasts:

> By the truth of my Lord, I have saved by it (i.e. the elixir) from this illness more than a thousand people and this became visible to all the people in only one day.
>
> (Kraus 1935: 303; translation RF)

Jābir here makes clear that the elixir that he owns is not only a panacea, but also a very rapid one: just like the transmutation is believed to take no time at all, so also was the cure of the sick achieved immediately. Jābir then goes on to describe how he treated a slave girl of the vizier Yaḥyā b. Khālid (see Strohmaier 2016): the slave girl had already been treated with a laxative to no effect, when Jābir was called for. As Jābir first was not allowed to see her, he prescribed cold water to be poured over her. But this did not help, nor did anything hot. When heated salt also did not help, Jābir was finally asked to see the slave girl in order to treat her more efficiently:

> And I saw a dying and very weak girl. I had a bit of this elixir with me, and gave her the weight of two grains with three ounces of pure oxymel [a honey–vinegar preparation] to drink. And by God and the truth of my Lord, I had to cover my face from this slave girl, because she returned to the most perfect state she ever was in in less than half an hour. Yaḥyā prostrated himself before my feet and kissed them and I said to him: O brother, don't! He begged for the use of the medicine and I told him: Take what I have with me of it. But he did not do so. Rather, he started to exercise and to study the sciences and similar things until he knew many things.
>
> (Kraus 1935: 304; translation RF)

Jābir also explained how he cured his own slave girl who suffered from mercury poisoning: after he had used all substances that he knew to be effective against poisons and nothing helped, he gave her the weight of one grain of the elixir with water and honey. As soon as she had drunk it, the girl started to throw up and was cured. Jābir remarked that the elixir in fact is the most effective theriac (i.e. antidote) there is. He then told the story of how he cured a man who had been bitten by a snake and was close to death with just two grains of the elixir dissolved in cold water (Kraus 1935: 305).

The Jābirian corpus was not alone in referring to the medical effects of the elixir. In his treatise on the categorization of the sciences, Ibn al-Akfānī wrote:

> The elixir of the stone causes different effects depending on the receptacle. It changes silver into gold, it dyes the white corundum red, makes mercury steady, and in medical operations, it has effects above those of medical drugs, and it frees from epilepsy, leprosy, elephantiasis and the like, as Ḥunayn b. Isḥāq has stipulated in one of his treatises in this intention.
>
> (ed. Witkam 1989: §§ 686–7)

This citation is very interesting not only because Ibn al-Akfānī stated clearly that at least some scholars were sure about the elixir's usefulness in medicine, as it was a panacea, but even more so for the fact that he cited the famous physician and translator Ḥunayn b. Isḥāq (d. 873 or 877). We must doubt that the untitled treatise mentioned by Ibn al-Akfānī really was by Ḥunayn, as we know him to have been an opponent of alchemy from the writings of al-Jildakī (Ullmann 1972: 249). However, the citation proves that beliefs in the medical effects of the elixir were quite prominent.

Alchemy as (natural) philosophy

In addition to its relation to medicine, alchemy must be also seen as a kind of natural philosophy. The alchemists themselves – or rather, the authors of alchemical works – often spoke of themselves as "philosophers" (see Todd 2016: 127). Accordingly, premodern critiques of alchemy usually do not blame alchemists for never actually producing gold, nor for being charlatans (as perhaps al-Bīrūnī does; see Carusi 2000: 461, n. 2), but rather they question the philosophical preconditions of transmutation.

Among those questioning the veracity of alchemy are people like the philosopher al-Kindī (fl. mid-ninth century) or the historian al-Masʿūdī (d. 956; see Ullmann 1972: 249–55). Ibn Sīnā's critique was based on his view that the alchemists had not truly mastered the Aristotelian theory of matter and species. He explained that transmutation is impossible because metals are not actually subgenres of one species, but different by specifics, and that it is impossible to change the specifics of any given substance, so alchemists would only manage to produce something that looked like gold or silver (Madkūr et al. 1964: 22–3; see also Carusi 2000: 461). As Ibn Sīnā's negative attitude toward alchemy was quite clear, the alchemical texts attributed to him should probably be regarded as spurious or not as texts on alchemy. The latter seems to hold true for the *Risālat al-iksīr* ("The epistle of the elixir"), which is not primarily on alchemy, but on the dyeing and alloying of metals (Anawati 1996: 878–9), though some passages seem to suggest that a partial transmutation could be seen as feasible (Moureau 2016a: 19–23).

Another prominent opponent of alchemy was ʿAbd al-Laṭīf al-Baghdādī (d. 1232), who in his youth had been an adept himself, but then became a philosopher. In his *Risāla fī mujādalat al-ḥakīmayn al-kīmiyāʾī wa-l-naẓarī* ("Epistle about the disputation of the two wise men, the alchemist and the theoretical philosopher," ed. Allemann 1988), he lets his former self, the alchemist, dispute with his current philosophical self; it is no surprise that the philosopher convinces the alchemist that the art is not a true one (Joosse 2008).

Prominent defenders of alchemy obviously include well-known practitioners of the art, like Abū Bakr al-Rāzī (d. 925 or 935), who defended alchemy against the accusations of the philosopher al-Kindī in his *Kitāb al-radd ʿalā l-Kindī fī raddihī ʿalā l-ṣināʿa* ("Book in refutation of al-Kindī's refutation of the art"). Of others, practical activities are not known, but authors like the geographer al-Hamdānī (d. 945?) follow the line that alchemy simply accelerates natural processes and should therefore be considered feasible (see also Todd 2016: 127).

The philosophical aspects of alchemy are sometimes so prominent that the alchemical aspect of a text does not become clear, as happened to the *Kitāb sidrat al-muntahā* ("Book of the zizyphus tree of the furthest boundary," ed. Braun 2016), attributed to Ibn Waḥshiyya (fl. ninth to tenth centuries), a work of allegorical alchemy that has been called a work on natural philosophy and philosophy of religion (Pertsch 1880: 375, no. 1162).

Critique and acceptance by religious scholars

The interplay of alchemy with religion can be seen on several levels, with some alchemists being religious scholars, some religious scholars and thinkers being considered alchemists, and some approving and some rejecting alchemy.

Among authors of alchemical texts, only very few can be identified as being religious scholars. Most prominent among these is the Andalusian Ḥadīth scholar Abū l-Qāsim Maslama al-Qurṭubī (d. 964), author of twin treatises entitled *Rutbat al-ḥakīm* ("The station of the wise") and *Ghāyat al-ḥakīm* ("The goal of the wise," the Latin *Picatrix*). In the *Rutbat al-ḥakīm*, alchemy was treated as the necessary last step toward mastering the highest level of the sage (i.e. magic, the subject of the *Ghāyat al-ḥakīm*). The alchemist, according to al-Qurṭubī, had to master mathematics and the natural sciences in order to be successful, and he also emphasized the practical aspects of alchemy as indispensable for the adept. That al-Qurṭubī really had practical experience in the laboratory becomes clear when reading his instructions for the purification of gold and silver (Moureau and Thomas 2016: esp. 107–8).

An interest in religion is also prominent in the collection of alchemical poems entitled *Shudhūr al-dhahab* ("The splinters of gold") by the twelfth-century Moroccan author Ibn Arfaʿ Raʾs. Until recently, this author was, on the basis of a passage in the historical work of Ibn al-ʿAbbār that has entered the later bio-bibliographical dictionaries, identified with a scholar of Qurʾānic variants, Ḥadīth, and Islamic law, bearing the same first name and the same patronym, but a different nickname (i.e. ʿAlī b. Mūsā Ibn al-Niqirāt). The identity of these two authors seems unlikely given that they are not identified by authors like Ibn Khaldūn or Ibn al-ʿArabī. Still, the author of the alchemical collection of poetry must have had a strong interest in religion and obviously was convinced that religion and alchemy were compatible, as his poems are strong in religious vocabulary, allusions to stories of the prophets, Qurʾānic vocabulary, etc. (see also Todd 2016: esp. 121–2). In one of his poems rhyming on the letter *ṭāʾ*, Ibn Arfaʿ Raʾs compares the alchemical process with the life of Moses, emphasizing the "transmutation" of the stick (e.g. see MS Berlin, Staatsbibliothek, Spr. 1969, fols 24v–5r). Similar passages may be found in the *Kitāb al-māʾ al-waraqī* by the Shīʿī author Ibn Umayl (Carusi 2000: 468–9). Qurʾānic citations are also prominent in the *Miftāḥ al-ḥikma*, attributed to a pupil of Apollonius of Tyana

(Carusi 2002), where several verses are used in order to defend the truth of alchemy (Carusi 2000: 465–8).

An interesting case is the Sunnī theologian and commentator of the Qur'ān, Fakhr al-Dīn al-Rāzī (d. 1210), who, though probably not a practitioner of alchemy, defended it, arguing yet again that alchemy is a simple imitation of natural processes, as metals grow in the earth just like an embryo in its mother's womb (Ullmann 1972: 253).

While evidence for an interest in alchemy from "orthodox" Sunnī scholars remains somewhat scarce, there seems to be a close interconnection of alchemy and Sufism and of alchemy and Shīʿī Islam. This is not surprising, as alchemy shares key concepts with Neoplatonism and gnosis. These are elements that are also important in the development of Sufism. Therefore, language and images of alchemical and Sufi writings resemble each other. Sufis tended to use alchemical vocabulary and metaphors to speak about God and the believer's quest, as is the case with Abū Ḥāmid al-Ghazālī (d. 1111), Ibn al-ʿArabī (1165–1240) and Jalāl al-Dīn Rūmī (1207–1273). For example, when an alchemist speaks of red sulfur, he might mean the elixir, while in Sufi writings, red sulfur is used to mean God. While the alchemist aims at the perfection of the metals, the Sufis were looking for the perfection of the soul. Accordingly, many famous Sufis were credited with writing alchemical treatises, as is the case with al-Ḥasan al-Baṣrī (d. 728), Sufyān al-Thawrī (d. 778), Dhū l-Nūn al-Miṣrī (d. 861), al-Junayd (d. 910), or al-Ghazālī (Ullmann 1972: 149, 196–7, 227; Addas 1993; Johnson 1996; Artun 2013: 21–2).

There seems to have existed a special affinity between alchemy and Shiite Islam: alchemical gnomologia are attributed to the first Shiite *imām* (and fourth Sunnī caliph), ʿAlī b. Abī Ṭālib, and the sixth *imām*, Jaʿfar al-Ṣādiq, is considered to be the author of several alchemical works and the teacher and master of Jābir b. Ḥayyān (Ullmann 1972: 195–6). In the Jābirian corpus, concepts of the Ismāʿīliyya have been detected, though they might not be Ismāʿīlī per se, but could perhaps be labeled proto-Shīʿī or seen as concepts of the extremist Shīʿa (*ghulāt*; Forster 2016: 19, 22).

When the elixir is compared to the *imām*, for example, as happens in the *Kitāb al-usṭuqus al-uss* ("Book of the element of foundation," see Lory 2016), one of the works of the Jābir corpus, this does not necessarily imply a Shīʿī milieu for the authorship, but it certainly helped the text to be read and accepted in just such a milieu. The Ismāʿīlī lineage is also prominent when the Ismāʿīlī missionary (*dāʿī*) Abū Yaʿqūb al-Sijistānī (executed shortly after 971) is considered to be the author of a book about the elixir, the *Kitāb al-gharīb fī maʿnā* (or *maʿrifat) al-iksīr* ("The book of the stranger on the meaning (or knowledge) of the elixir," see De Smet 2003). Even the famous al-Ḥākim (d. 1020), Fatimid caliph and founding figure for the Druze religion, is credited with the authorship of a treatise on alchemy (Brockelmann 1937–1949: S I: 902; Storey 1977: 435–6, no. 757).

Although skepticism toward alchemy from religious scholars was much less pronounced than it was with other occult sciences, namely astrology (Ullmann 1972: 249–55 and 274–7), some religious scholars were outspoken critics of alchemy. The most prominent among them is the radical conservative jurist Ibn Taymiyya (d. 1328). He appears to have bought books on alchemy from the estate of a deceased preacher (*khaṭīb*) and to have washed the pages clear, arguing that the text these pages had contained had led people astray (Ibn Ḥajar al-ʿAsqalānī n.d.: III: 39; see also al-Ḥaẓīmī 2003: 60).

As discussed above, religion and alchemy are strongly interlinked, with religious scholars both defending and attacking alchemy. As there seems to have been a special interest in alchemical practices among Sufis, it might be considered a logical consequence that the jurist Muḥammad b. Muḥammad Ibn al-Ḥājj (d. 1336; see Brockelmann 1937–1949: S II: 95), a Moroccan, dedicated a whole section of his work *al-Mudkhal* ("The introduction") to alchemy: in that section, he reminds any Sufi adept that striving for the secrets of alchemy is wrong, as the true Sufi should be striving for the other world, and that alchemy would lead to nothing but poverty. He admits that gold and silver are efficacious if used in medicaments, but insists that to that end, normal gold and silver, obtained by mining, are sufficient, and that alchemy was really nothing but a waste of money (Ibn al-Ḥājj n.d.: esp. 138, 144). As *al-Mudkhal* is addressed to a Sufi audience, we might deduce that there was, indeed, some serious interest in alchemy in Sufi circles, as the section otherwise would have been quite superfluous.[6]

LATIN WORLD

Jean-Marc Mandosio

Categorization of the sciences

When the first translations of Arabic alchemical texts were made available during the twelfth century, Western European scholars discovered the existence of an art that, paradoxically, seemed both new and old. It was entirely new to them, for no such discipline had ever been mentioned in the Latin scientific tradition; but it presented itself as a very ancient lore. The putative earliest Latin introductory text on alchemy, supposedly written in 1144 and ascribed to Robert of Chester (Lemay 1991; Moureau 2020), assumes that this art is unknown to the readers and explains that it was revealed by Hermes Trismegistus "after the Flood." The categorization of the newly adopted discipline in the Latin "divisions of philosophy" – that is, classifications of the sciences – did not follow Arabic models; the dominant scholastic conception was to view alchemy as a "mechanical art" (Mandosio 1991: 201–10).

The first attempt to categorize alchemy goes back to the Spanish cleric Dominicus Gundissalinus (d. after 1181). His work *On the Division of Philosophy* (*De divisione philosophiae*, ca. 1150, ed. Fidora and Werner 2007) is largely based on al-Fārābī's *Enumeration of the Sciences*, a work that he translated and that did not mention alchemy; to compensate for the absence of an art that had already begun to raise the interest of learned Latin readers, he used the section dedicated to alchemy in another Arabic treatise, *On the Origin of Sciences* (*De ortu scientiarum*, ed. Bauemker 1916), falsely ascribed to al-Fārābī. Gundissalinus places alchemy – together with medicine, agriculture, navigation, catoptrics, and several magical disciplines – among the "particular" sciences, aimed at practical operations, whose principles are grounded in the "universal" natural science. Natural science is "the science which brings forth the principles of natural bodies and their accidents," while alchemy is "the science of the conversion of things into other species."

Elaborating upon Gundissalinus' stance thanks to the scholastic concept of "subalternation" of the sciences, Thomas Aquinas (d. 1274), in his commentary of Boethius' *De trinitate* (before 1259, ed. Fratres Prædicatores 1992), places alchemy, along with medicine, agriculture, "and all other like sciences," among the disciplines that are subordinated to natural philosophy. He dissents from Gundissalinus on a key point: alchemy cannot be considered as a "species" of natural philosophy, as Gundissalinus asserted, for it belongs to a group of "inferior" sciences that operate upon a specific kind of body (minerals in the case of alchemy), while the "superior" natural science abstractly determines the general principles regarding all bodies. Its subaltern status makes alchemy one of the mechanical arts, as another Dominican, Robert Kilwardby (d. 1279), explains in his treatise *On the Origin of Sciences* (*De ortu scientiarum*, ca. 1250, ed. Judy 1976): "the speculative science is by essence apt to examine the causes" of things – it therefore addresses the question *propter quid* (meaning "why") – while "the mechanical art only deals with bodily operations and works" and cannot rise above the question *quia* (meaning "how").

The standard list of seven mechanical arts, mirroring the seven liberal arts, was devised during the first half of the twelfth century, when alchemy was still unknown to Latin scholars, in the Parisian school of St Victor. Famously discussed by Hugh of St Victor (d. 1141) in his *Handbook on Reading* (*Didascalicon de studio legendi*, ed. Buttimer 1939), it comprises weaving, equipment (including blacksmithing and architecture), navigation (i.e. trade), agriculture, hunting (including cooking), medicine, and the theatrical arts. The same list was reproduced by Richard of St Victor (d. 1173) in his *Book of Excerpts* (*Liber exceptionum*, ed. Chatillon 1958). One century later, drawing on Richard's enumeration to frame his own description of the mechanical arts, the Dominican Vincent of Beauvais (d. 1264) acknowledged in his *Doctrinal Mirror* (*Speculum doctrinale*, ca. 1250; ed. 1624) the existence of alchemy by inserting it in place of medicine. The latter, Vincent asserts, cannot be categorized as a mechanical art for it is not merely practical but also has a theoretical dimension – hence its status as a doctoral discipline in recently founded universities. Alchemy, for its part, "can rightly be counted among the mechanical arts" – a statement that implies that the theoretical aspects of alchemy are not essential. As an art aimed at dividing inanimate substances into their component parts (as Vincent defines it), alchemy is useful to metalworking "because it examines the nature, mixture, division and transmutation" of metals, and to medicine because it deals with "the separation of healthy substances or qualities from the harmful ones which are often mixed with them."

It should be noted that the categorization of alchemy as a mechanical art was an external view that did not take into account the philosophical claims of the alchemists themselves. This is why Vincent of Beauvais, being a compiler who wishes to provide his readers with the most comprehensive information available, stresses that alchemy is also defined by some as a philosophy. In so doing, he takes into account both the scholastic and the alchemical conceptions of this discipline.

Alchemy and other occult sciences

The medieval alchemists were eager to assert themselves as "natural philosophers" and not to be confused with magicians or sorcerers. Therefore, their writings were seldom explicitly related to other "occult" activities. Even though the names used to designate metals in alchemy were those of the seven planets, astrology played little or no role in the completion of the Great Work – with the exception of several notable instances such as the astrological determination of the best time to operate, mentioned in the Jābirian *Book of the Seventy* (*Liber de septuaginta*) and in the pseudo-Avicennan alchemical treatise *On the Soul* (*De anima*, ed. Moureau 2016b). However, alchemy could be – and sometimes actually was – considered as a component of a wider area of knowledge dedicated to occult phenomena. One of the definitions of alchemy reported by Vincent of Beauvais described it as an "occult philosophy." This conception was supported by *Picatrix* (i.e. the Latin translation of Maslama b. Qāsim al-Qurṭubī's *The Goal of the Wise* [*Ghāyat al-ḥakīm*] made from a lost Spanish version in the second half of the thirteenth century [ed. Pingree 1986]). According to this work, alchemy is one of the four parts of "nigromancy": theoretical nigromancy deals with the knowledge of the stars and their influxes and with the knowledge of names and their powers, while practical nigromancy deals with the fabrication of talismans (defined as the action of "a spirit in a body") and with alchemical operations (the action of "a body in a body"). Though considered unfit for Christians, this treatise had an underground influence that is difficult to ascertain. A faint echo of it can be traced in the first book *On Natural Magic* (*De magia naturali*, ed. Mandosio 2018b), written in 1492–1494 by the French philosopher Jacques Lefèvre d'Étaples (d. 1537), who claims that the earliest practitioners of magic, the Chaldaeans, "were either astrologers, physicians, transmutators, or all these together." Moreover, the conception of quintessence, formulated in the fourteenth century, established an actual link between the stars, alchemy, and medicine. A good case in point is the Dutch physician Peter of Zealand, who wrote a small treatise on alchemy (*Opus Petri de Silento*, ed. Zetzner 1613: 4: 985–97), long ascribed to an elusive "Petrus de Silento" (a misreading of "Petrus de Zelante" or "Zelandia"), and also a lengthy work on magic, the *Elucidation of Marvelous Things* (*Lucidarius de rebus mirabilibus*, 1491–1494; Mandosio 2019), in which astral rays play a fundamental role and quintessence is described as the best medicine there is, for it is imbued with a "formal" (i.e. celestial) power incomparably more powerful than any "material" drug.

Alchemy as philosophy

The "mechanical" label currently attached to alchemy in the scholastic context was at odds with the alchemists' promotion of their own art as the crowning achievement of natural philosophy. The great defender of this concept was the Franciscan Roger Bacon (d. 1294; Newman 1997). In his *Opus tertium* (1267, ed. Brewer 1859), he granted alchemy a twofold nature, going so far as to divide it into two distinct sciences: one practical, the other speculative. Operative alchemy teaches "the artificial production of noble metals, colours and other things, better and more abundantly than Nature does," while speculative alchemy deals with "all things inanimate." Bacon considered the latter as the key to the true understanding of natural philosophy and medicine. However, reversing the usual scholastic subordination of "mechanical" practice to theory, Bacon asserted that operative alchemy "certifies" speculative alchemy, and he considered it as the foundation of the "experimental science" – the purveyor of unseen marvels that he called for. According to his views, operative alchemy has no need of any "higher" theoretical discipline to be grounded in, and therefore should not be categorized as a subordinate "mechanical" art. With such a high appraisal of their art, it is no wonder that Bacon became an authority for later alchemists, notwithstanding the fact that most, if not all, alchemical works ascribed to him were apocryphal.

Of course, to consider alchemy as a philosophical discipline implied the acceptance of its transmutational claims, on the grounds that a science dealing with nonexistent things is not a science (e.g. there is no science of the void). If what the alchemists say is true, alchemy is a science; if not, it is an *ars sine arte* ("art without art"). Therefore, the identification of alchemy with the highest kind of philosophy could be made only by the alchemists themselves and their followers. The earliest Latin commentary of Hermes Trismegistus' *Emerald Tablet*, called *Explanation of the Words of Hermes, Master of Philosophers, According to Our Truth* (*Expositio verborum Hermetis magistri philosophorum secundum veritatem nostram*, twelfth century, ed. Steele and Singer 1928), accompanied the "vulgate" (*vulgata*; i.e. most common, widespread) translation of the *Tablet*, which interpreted the name *Trismegistos* ("thrice-great") as meaning "the three parts of philosophy" (i.e. ethics, logics, and physics; Mandosio 2004), implying in turn that the secrets of alchemy could be grasped only by the most consummate philosophers.

In accordance with this view, Pseudo-Geber's *Compendium of the Achievement of the Great Work* (*Summa perfectionis magisterii*, late thirteenth century, ed. Newman 1991) claims that the alchemist "should be learned and advanced in the knowledge of natural philosophy." The very form of the work, modeled on the scholastic *summae*, suggests that alchemy has an intellectual dignity to which its purported "mechanical" status does not give justice. The twofold structure of many alchemical treatises (and of the *Summa perfectionis* itself), divided into "Theorica" and "Practica," also tended to demonstrate that Vincent of Beauvais' argument aimed at differentiating alchemy from medicine, based upon the assumption that alchemy was in essence a practical discipline, had no legitimate ground, since this art possessed, just like medicine, its own theoretical part. The ponderous *New Precious Pearl* (*Pretiosa margarita novella*, 1330, ed. Manget 1702: 2:1–80), written by the Italian physician Pietro Bono of Ferrara, is a further step in the same direction.

Critique and acceptance by scholars

There was no consensus among medieval scholars on the possibility of alchemical transmutations. Some, like Roger Bacon, believed that the alchemists' claims were true; others, such as Ramon Llull (d. 1315), believed that they were false and that alchemists were either frauds or deluded; still others, most notably Thomas Aquinas, believed that the transmutations they bragged about were conceivable theoretically but very unlikely to be put into practice.

This last stance was derived from Avicenna's argument against transmutations, which had acquired an authoritative status among the Scholastics due to its early attribution to Aristotle. The chapters of Avicenna's *Book of the Healing* concerning minerals were excerpted and translated into Latin by Alfred the Englishman (d. after 1222) in the last decades of the twelfth century, in order to be inserted at the end of the Latin version of Aristotle's *Meteorologica* so as to make up for the lack of a text *On Minerals*, which Aristotle was thought to have written (*De mineralibus*, ed. Rubino-Pagani 2016; Mandosio 2018a). As a result, the objection raised by Avicenna passed for a statement made by Aristotle, giving it a much higher status in the Latin world than anywhere else, even though Avicenna's prestige was far from negligible to begin with.

Here is Avicenna's objection as it is phrased in Alfred's Latin version:

> The artificers of alchemy should know that they cannot change metallic species, although they can make imitations. Why is it so? Because a metallic

> compound cannot be changed into another compound, unless perhaps it be brought back to its prime matter, so that it would be changed into something else than what it was before. Otherwise, I do not believe that it is possible to remove the specific difference by any contrivance.

It is interesting to note that the Aristotelian concept of "prime matter" did not explicitly appear in Avicenna's original wording (Holmyard-Mandeville 1927) and was added by the translator, making the text appear as if it was really written by Aristotle.

The ascription of Avicenna's objection to Aristotle had a twofold consequence: it was taken very seriously by the alchemists themselves, and its discussion, the so-called *quaestio de alchimia*, became a standard feature of the scholastic commentaries on Aristotle's *Meteorologica* (Newman 1989; see also Obrist 1996). This state of affairs continued even after it was made clear – first by Albert the Great (d. 1280) in the 1250s and a decade later with the release of the new translation of Aristotle's work from the Greek by William of Moerbeke (d. 1286) – that the infamous passage on alchemy was not genuine.

In fact, many alchemists adopted Avicenna's argument as their own, and going back to the prime matter of metals in order to produce transmutations became the generally accepted notion. All metals were seen by the alchemists as different states, from the "basest" (such as lead or iron) to the "purest" (silver and gold), of a single species – metal – and therefore the problem raised by the "specific difference" or "specific form" of each metal vanished, at least from a theoretical point of view. Furthermore, a rhetorical trick aimed at neutralizing the recurrent suspicion of fraud was to claim that such a suspicion was entirely justified, granted that there were two kinds of alchemists: the false ones, be they forgers or ignoramuses, and the "true philosophers," who were so rare that the probability of meeting a charlatan greatly exceeded that of meeting a real alchemist. This view was supported by Bacon, who claimed in the *Opus tertium* (ed. Brewer 1859) that he knew "only one man who can be lauded in the works of this science" (he did not say who the man was). The common conception, shared by the alchemists, was that the near impossibility of understanding the cryptic writings of the ancients and the extreme difficulty of the practice of the work itself explained why so few alchemists could succeed – to the point that Petrus Bonus asserted that the Great Work could not be achieved without God's help, hence the moral virtues necessary to obtain this divine assistance.

Alchemy and medicine

Antoine Calvet

The *Liber compostelle* of Bonaventura da Iseo, composed between 1253 and 1268, contains a *Tractatus aquae vitae*, one of the first distillation writings –

that of wine – accompanied by a practical alchemical text (Calvet 2018a: 130–3). In the medical field, distillation was introduced to Europe by the physicians Taddeo Alderotti (1295) and Teodorico Borgognoni (1205–1298). The texts on distillates, the latter advertised as panaceas, were received with reluctance by some of the greatest professors of medicine, such as Arnald of Villanova (1311; Calvet 2011), but praised and acclaimed by others such as Bernard de Gordon (fl. 1293–1308). In contrast to the image he left behind – that of having been a confirmed alchemist – Master Arnald spoke out forcefully against the use of alchemico-medical products such as the elixir of metals, obtained as a result of alchemical operations (Crisciani and Ferrari 2010). As for *aqua ardens* (alcohol), he sometimes praised its subtlety, without attributing to it the wonderful effects that distillation texts attributed to this product. In any case, very quickly important writings of distillatory alchemy such as the *De aqua vitæ simplici et composita* circulated in the fourteenth and fifteenth centuries under his name.

The relationship between medicine and alchemy, at least in the Latin world, was established and strengthened toward the end of the thirteenth century and the beginning of the fourteenth century. The reading of the *Canon* of Avicenna is a good testimony to the spread of alchemical distillation in medical treatises. Avicenna saw in the decanting of blood, carried out in a glass vessel, proof that it was a compound of the four elements (*Canon* I, 1.4.1). The Latins understood this decanting vessel as a *distillatorium*. Thus, Roger Bacon used this passage to reinforce his theory of distilled blood, by which life could be extended. Similarly, Taddeo Alderotti felt that, in this extract from the *Canon*, Avicenna alluded to an alchemical practice. On the other hand, Gentile da Foligno (1348), famous for his commentary on the *Canon*, refused to believe in the utility of any alchemical distillation (Chandelier 2017: 389–90). Whatever the resistance of physicians to integrate the products of distillation (elixir, *aqua ardens*, drinkable gold) into their therapeutics, medical alchemy was a great success. In the form of health plans attributed to Arnald of Villanova, Ramon Llull, or Roger Bacon highlighting the virtues of the philosophers' stone, the idea spread that gold or blood passed through the still supplanted all medicines and was valuable for all diseases, such that it had an effect on every temperament (Paravicini Bagliani 2003). At a time when the plague was afflicting everybody indiscriminately, these medicines seemed to be an obvious tool to use in the epidemic (Crisciani and Pereira 1998). Was it a coincidence that the author of a remarkable treatise on the alchemical quintessence, Jean de Roquetaillade (ca. 1356), was a survivor of the plague that killed many brothers of the convent of Rieux (Haute-Garonne), where he was confined?

Drinking gold was also associated with obtaining a longer life, which seems to have been an obsession of several popes (John XXI, Clement V, Boniface VIII, and John XXII); some of these alchemical treatises are also addressed to them. In the fifteenth century, the popularity of distilled water and drinking gold led to a new religious order: the Jesuati. They followed the ideas of the *Devotio moderna* and were influenced by the thought of Henri Suso, giving the mixture to the poor. They were given the name *fratelli dell'aqua vita* (Calvet 2018a: 175–6). The order was suppressed by Pope Clement IX in 1668.

ALCHEMY IN THE JEWISH CONTEXT

Gabriele Ferrario

Alchemy was a thriving discipline, an object of scholarly discussion, and, for some, a path to promised wealth in the medieval societies where Jewish communities were present during the Middle Ages, in both the Islamic and Christian contexts. There is no doubt that Jewish scholars were aware of the debates regarding the validity of alchemy, and Jewish people were attracted by its practice. Nevertheless, the relevance of alchemy, its status in the Jewish milieu, and its relationship with Judaism and with the other philosophical traditions and practical arts appear to have raised questions that ancient and medieval authors asked themselves as much as modern scholars have done and do, reaching often different conclusion.[7] The scarcity of original alchemical treatises written by Jews during the Middle Ages appears surprising in particular since Jewish figures – real or fictional – are mentioned by non-Jewish alchemists in the earliest alchemical manuscripts as originators of alchemical knowledge and are often referred to as teachers of important secrets and practices throughout the history of alchemy. For example, the first historical figure of an alchemist, Zosimos of Panopolis, who flourished around 300 CE, shows deep reverence to and often quotes his teacher, Mary the Jewess, a rare female figure of an alchemist suspended between historicity and fiction: Mary was variously identified as Jewish, a Copt, a Prophet (*Maria Prophetissa*), the wife of Moses, and the sister of Jesus.[8] More generally, patriarchs, prophets, and characters found in the Hebrew Bible and in apocryphal literature are invested with alchemical knowledge and teachings in a number of late antique and medieval sources. This was not always considered a distinction; rather, the *topos* of the Jewish alchemist served a doubly derogatory function: to discredit the Jewish people as propagators and practitioners of a fraudulent discipline (Ferrario 2010). Modern scholars have assessed the relevance of alchemy in the Jewish context in different (and often opposite) ways. Moritz Steinschneider (1873) strongly repudiated any Jewish interest in alchemy because of the danger of the accusation of counterfeiting that may come with it and its fundamental irrationality, and Gershom Scholem (1925), while presenting examples of alchemical writings produced and circulating in the Jewish milieu, showed the radical incompatibility of the basic principles of Jewish kabbalah and alchemy. There are only few mentions of the transmutation of metals and of other alchemical themes in medieval Hebrew mystical writings (Bos 1994); the association of kabbalah with alchemy picked up in the early modern period, mostly fostered by the circles of Christian kabbalists (on Christian Kabbalah, see Forshaw 2013). Recently, in an effort to reevaluate the Jewish involvement in alchemy, Rafael Patai (1994) produced a substantial monograph aimed at showing the long-lasting and numerous involvements of the Jewish people in alchemy, but his methods and conclusions have received criticism. The scarcity of original alchemical treatises penned by Jews during the Middle Ages, in comparison with the abundance of coeval Arabic and Latin material, was rightly underlined by Gad Freudenthal (2011), who recently authored an overview of the available evidence of these writings and characterized the Jewish contribution to medieval alchemy as a "noted absence."[9] While trying to evaluate the diffusion of alchemy in the medieval Jewish milieu, it must be kept in mind that alchemy was not a purely intellectual enterprise but, as a medieval art, had an inherent and important practical component: although material traces of the practice of alchemy in the Middle Ages are scarce, we may find some indications of it in recipes and collection of recipes that appear to be devoid of theoretical and philosophical aspects and may have been part of the library of practicing alchemists. Court records and administrative documents may also reveal details on the lives of alchemist who did not leave any original written work. Moreover, the survival rate of medieval manuscripts allows us only a partial view, and this may be truer in the case of a discipline that does not always find favor in the eyes of religious or secular authorities. In what follows, I am going to present a brief overview of the extant evidence, which shows a varying degree of acceptance of alchemy in the Jewish medieval context: from strong criticism to enthusiastic approval, from avoidance to active involvement.[10]

Alchemy is presented as a dangerous practice in a record written in the rabbinical court of Fusṭāṭ in the tenth century (Scholem 1925: 8; Assaf 1942: 115; Freudenthal 2011: 344), where a fraudulent alchemist managed to flee after being entrusted with some precious stones, leaving the deprived owner to the punishment inflicted by the authorities who suspected his association with the fraudster.[11] The recorder of the case commented that alchemy was "a practice proper to kings," and added that "whoever approaches it, destroys himself with his own hands."

The evidence retrieved from the Cairo Genizah shows the existence of an interest in alchemy among the Jews of the medieval Mediterranean world. The importance of the material deriving from the Cairo Genizah – a storage room for discarded sacred and secular writings attached to the Ben Ezra synagogue of Fusṭāṭ, Old Cairo – cannot be overstated (see Goitein 1967–1993). A relatively small corpus of fragments, which may preserve only a few lines of text or be constituted by groups of bifolia deriving from booklets and anything in between, portrays a significant involvement of local Jews in both alchemical theory and its practice.[12] The fragments are generally in Judeo-Arabic (Arabic written in Hebrew script), the language commonly used for secular writings in the Jewish communities of Arabic-speaking countries, with a few Hebrew exceptions. There is also a relatively small group of fragments written with the Arabic alphabet that could point to the Arabic sources for the Judeo-Arabic tradition. The vast majority of the Genizah alchemical corpus is made up of single alchemical recipes and collections of recipes aimed at metallic transmutation and, more rarely, at the production of precious stones; they are transmitted anonymously and are devoid of the very common hermetic strategies of alchemical texts. The use of *Decknamen* ("cover names"), for example, is limited to the substitution of the names of metals with those of the corresponding planets, and this feature does not appear with any consistency in the fragments. These features assimilate these recipes to the operative alchemical traditions transmitted in Greek papyri, Byzantine alchemical collections and their Syriac counterparts, and numerous treatises and collections produced in the Arabo-Islamic world. Rarer, but very telling of the interests of the Jews of medieval Egypt, are the few leaves transmitting titles of known alchemical works, names of alchemical authorities, or identifiable alchemical theories. A survey of the corpus has shown that the names of Jābir b. Ḥayyān, Ibn Umayl al-Tamīmī, Ibn Waḥshiyya, Khālid b. Yazīd, the Pseudo-Aristoteles and Hermes,[13] together with the titles of some of their works, were known and read by the Jewish alchemists that left their traces in the Genizah. The identity and social position of those involved in the transmission of alchemical texts and in alchemical practice in medieval Cairo are still matters of debate: the extant evidence does not allow any clear assessment, and their identification with the circles of physicians and druggist among the Jewish communities is to be considered only as a plausible hypothesis.[14] Genizah manuscripts witness an interest in the theoretical and practical aspects of alchemy among Jews living in the Islamic world, an interest that appears to be still present up to modern times (Yinon 1993: 93; Patai 1994: 370–1).

The *Emm Ha-Melekh* ("The Mother of the King"), an Arabic alchemical treatise whose original is now lost, was translated into Hebrew in thirteenth-century Provence and gained a certain popularity (Scholem 1926; Patai 1994: 98–118). Its aim was to persuade the reader of the validity of alchemy and the truthfulness of its promises. In particular, the *Emm Ha-Melekh* argues against the philosophers who maintain that alchemical transmutation would imply the impossible change in the species of metals by stating that alchemists perform in their laboratories the same kind of "maturation" that happens naturally in caves over a long time and that no *differentia specifica* is thus created in their operations. The *Emm Ha-Melekh* not only defends alchemy's validity, but also offers descriptions of alchemical operations and provides three recipes that are said to be derived from ancient authors. In the late medieval period, portions of the *Emm Ha-Melekh* appear as interpolations in a pseudepigraphic

Maimonidean epistle, the *Iggeret Ha-Sodot* ("Epistle of Secrets"), which is very likely a product of the Jewish milieu of Andalusia after the *Reconquista* (Patai 1994: 309–13; Freudenthal 2011: 349).

A recently discovered Hebrew translation of an alchemical treatise is possibly the only witness of the transmission of alchemy from a vernacular (Catalan in this case) into Hebrew. The text is devoted to practical instructions on the production of the alchemical stone and is scribbled in a very difficult hand on the margins of a Greek Euclidean manuscript preserved in Toledo's cathedral: the problematic language and its marginal layout in the manuscript led scholars to consider it an autograph produced for the personal use of its author (Freudenthal 2011: 351–2).[15]

Evidence shows that in Christian lands, during the fourteenth century, the kings of Aragon employed at different times two Jewish alchemists (Patai 1994: 234–7).[16] In 1345, the first of the two, known as Magister Menaḥem, was accused of necromancy, counterfeiting, and "experimenting" in Palma de Majorca and was later employed as a personal physician and alchemist by King Pedro IV. The second, Caracosa Samuel of Perpignan, was granted by King Juan I the right to practice alchemy undisturbed by the threat of persecution from local authorities. Jewish sources reveal that Caracosa Samuel was a well-seen member of the local Jewish community, but they never mention the nature of his practices. A handful of similar references to Jewish alchemists providing their services to rulers and notables in Central Europe, Northern Italy, and Southern France is found in the fifteenth century: alchemy is often associated with the counterfeiting of money and, on more than one occasion, these Jewish alchemists received such accusation and were jailed and executed (Mentgen 2009; Freudenthal 2011: 355).

European libraries preserve few very interesting Hebrew miscellaneous manuscripts of alchemical content,[17] and since they are all dated between the sixteenth and the seventeenth centuries, they fall outside the chronological span of this survey. Nevertheless, these manuscripts transmit important medieval Arabic and Latin works in Hebrew translation and may be the final ring in a chain of transmission that begun in medieval times. For example, a pocket-sized manuscript in the Berlin Staatsbibliothek preserves, among anonymous and unidentified treatises, Hebrew translations of influential Arabic works, like the famous *Book on Alums and Salts* pseudepigraphically attributed to Muḥammad ibn Zakariyyā' al-Rāzī; it is very likely that this work was composed in Arabic in twelfth-century al-Andalus and translated into Latin shortly afterwards (Ferrario 2007).

How was alchemy considered among Jewish philosophers and scholars? They surely were aware of the ongoing debates on the validity of alchemy among Muslim and, later, Christian scholars. Alchemy is not central to their arguments but appears in passing mentions, often as a term of comparison that may reveal the authors' perception of the discipline and tell us something about its consideration and position among medieval sciences. It must be noted that the works of these scholars were circulating in Hebrew and spread among Jews in non-Arabic-speaking lands, thus possibly influencing their European readers.[18] Judah Ha-Levi in the late eleventh century argued that alchemy had no theoretical basis; Baḥyah ibn Paqudah in the early twelfth century appears to have considered alchemy as valid (but still less effective than faith); Maimonides toward the end of the twelfth century criticized alchemical writings as confusing; and Gershon ben Solomon in the thirteenth century, while reproducing in his *Sha'ar ha-Shamayim* ("The Door of the Heaven") an entire section of the theoretical part the *Emm ha-Melekh*, appears to have distanced himself from the possibility of transmutation underpinning the work.[19] A certain interest in alchemy is shown in the early fourteenth-century Hebrew version of al-Fārābī's *Iḥṣā' al-'Ulūm* ("Enumeration of the Sciences"), whose translator, Qalonymos ben Qalonymos ben Meir, included a description of the methods and aims of alchemy, which is absent in the Arabic manuscripts of the Fārābian work, and was in possession of an Arabic copy of a Jābirian text that he found in the library of Robert of Anjou (Freudenthal 2011: 350).

As this brief survey has shown, medieval Jewish cultures had a peculiar relationship with alchemy, a discipline that did not attract large original production as was the case in the Islamic and Christian contexts nor was dealt with as extensively in non strictly alchemical Jewish sources. It is also plausible that Jewish authors chose to write their alchemical works in Arabic language: these works could be among the very numerous Arabic alchemical texts that were transmitted anonymously or pseudepigraphically attributed to famous ancient or Muslim alchemists. Among medieval Jewish scholars there was surely knowledge and interest in alchemy and in the natural philosophical problems that its aims and methods raised; alchemy was also practiced by Jewish individuals, who derived their knowledge from Arabic and Latin sources. Their practices have left important traces in the recipes preserved in the fragments from the Cairo Genizah, and, later on, their at times problematic relationships with rulers, notables, and local authorities are recorded in court proceedings and chronicles.

CHAPTER FIVE

Society and Environment: *The Social Position of the Alchemist and Alchemy in the Court, in the Church, and in Society*

CHARLES BURNETT WITH ANTOINE CALVET
AND JUSTINE BAYLEY

This chapter treats the position that alchemy and other chemical procedures held in society, both from the point of view of the alchemist and from that of the external observer. Among the categories into which these examples fall are the benefits and the dangers of alchemy to society as a whole and to the individual, the laws allowing and regulating alchemy, and the employment of alchemists by popes and secular rulers.

THE STATUS OF ALCHEMY AND THE ALCHEMIST

In what is probably the earliest translation from Arabic into Latin of an alchemical work, the *Liber Morieni*, completed by Robert of Chester on February 11, 1144, we read:

> Since you who belong to Latin culture have not yet understood what *alchymia* is and of what it is composed, I shall explain that here. ... Hermes the

> philosopher and others who came after him, defined the word in such a way: *alchymia* is a corporeal substance composed from one thing and through one [thing], joining the most precious things together through relationship and effect, and naturally converting the same things by a natural commixture and by the best artifices.
>
> (Ruska 1928: 30)

Here, "*alchymia*" refers to the *elixir*, rather than of the art of alchemy. Aside from "*Latinitas*" (Latin culture or Latin letters), we do not know precisely for whom Robert was composing his translation. His other translations, on astronomical tables, algebra, and astral magic, suggest that the aim was to introduce practical sciences. It is perhaps not a coincidence that the science of "alchemy" is mentioned by Dominicus Gundissalinus, an archdeacon of the diocese of Segovia (where Robert was also active), as the last of the practical sciences that are an offshoot from physics: medicine, astrological judgments, necromancy according to physics, talismans, agriculture, navigation, optics, and alchemy (see Chapter 4).

Robert would also have been working in an environment where astrological texts were being translated into Latin, in which the different professions were clearly distinguished from each other, and the alchemists are shown to have their own positions in society. Abū Ma'shar's *Book on Religions and Dynasties*, originally written in Baghdad in the mid-ninth century and translated into Latin probably in the mid-twelfth century in Toledo by John of Seville with the title *On the Great Conjunctions* (*De magnis coniunctionibus*), ranks alchemy alongside minting dirhams and copper fils ("silver or copper coins" in the Latin translation) and alchemists alongside others operating with fire, such as makers of weapons, adding that kings are likely to employ them as skilled craftsmen (like jewelers and weapon-makers), but that they work in secret (Yamamoto and Burnett 2000: bk III, lines 156–62). Another passage in the same work, however, associates the "easy" (uncontrolled) practice of alchemy with other symptoms of an unstable society: secret dealings, bad relations between the rulers and his subjects, and people fleeing from the society (Yamamoto and Burnett 2000: book III, line 181). Other astrological works translated from Arabic in the same period reflect the uncertainties about whether the alchemist is genuine or a charlatan but, on the other hand, advise on when to practice alchemy to gain the best results (Burnett 1992). In Arabic, one finds a whole chapter of al-Jawbarī's *Mukhtār fī kashf al-asrār* dedicated to the tricks of false alchemists (ca. 1200; Abrahams 1984).

The concerns about the alchemist in society in these Arabic astrological works continued to be experienced in "Latin culture" (*latinitas*). For Ramon Llull, alchemists are associated with the "changeable" planet Mercury. Mercury cannot be trusted as the alchemist's helper, for he is "a deceiver of alchemists, since he empties their purses and makes them wear torn coats" (Lullus, *Liber de regionibus sanitatis et informitatis*, d. 2, l; Gayà Esterlich 1995: 293–4).[1]

We meet these alchemists in the flesh, as it were, through other accounts. Leo Africanus, in the early years of the sixteenth century, tells us of an "infinite number of alchemists" in Fez, "stinking of sulfur and other things no less unpleasant to smell," and "meeting together every evening at the chief mosque, to discuss the contents of their fantastic imaginations" (Rodríguez Guerrero 2010: 297–8, 304–5). He characterizes them as "ignorant people, of a coarse kind" (Africanus 1830: I:421).

The author of the Pseudo-Albertus *Semita recta*, an alchemical work, on the other hand, reports that he has found "many very rich lettered people – abbots, provosts, canons, doctors, as well as illiterate people – who have spent a great amount on this art" (Halleux 1979: 120). The author of the *Summa perfectionis* of Pseudo-Geber describes the ideal prospective alchemist as someone who "must be fit, his soul must be midway between stationary and mobile, and he must have money" (Newman 1991: 635–8). But unsuccessful alchemists are poor because they spend all their money (Pseudo-Avicenna, *De anima*; Moureau 2016b: 621–3). In the early sixteenth century, alchemists could still be satirized for their poverty, as we read in a collection of fictitious letters written by the Christian cabbalist Reuchlin: "Just as now, every alchemist is (purports to be) a doctor or a masseur (*saponista*), who has arrived at hard times" (Boemer 1924: I:44).

Maslama b. Qāsim al-Qurṭubī (early tenth century) begins his work *The Rank of the Wiseman* (*Rutbat al-Ḥakīm*) by describing geometry, astronomy, logic, and the peripatetic natural sciences as the foundation for alchemy, and the scholar must have mastered these subjects before he reaches the requisite rank (*rutba*) for studying alchemy. Maslama describes the spiritual and intellectual progress of the wise man in terms of a ladder, whose final steps (*martabāt*) are alchemy and magic. The same language reappears in Latin in the *Opus maius* of Roger Bacon, in the template for a new curriculum of learning that he sent to Pope Clement IV in 1267, in which once again alchemy appears as the culmination of intellectual study. For Roger, alchemy supersedes the other arts and sciences because it adds practice to theory (Bridges 1897: II:214), a theme that we shall see recurring throughout this chapter. The implied society for these texts is a scholarly community. But it is important to note that alchemy never became one of the subjects studied at European universities.

There were dangers in practicing alchemy. In 1485, many houses in Vienna were burned when, on the occasion of Matthias Corvinus' capture of the city, several alchemists of the town celebrated with fireworks (Ogrinc 1980: 123). As we have seen, the company of alchemists in Fez was avoided because they stank of sulfur. Chaucer, in the *Canon's Yeoman's Tale*, refers to the consequences of the explosion of an alchemist's pot probably as a result of adding crushed litharge and saltpeter to the mixture (Gabrovsky 2015: 22–3). Sometimes the ingredients of alchemy came from stinking places. Pseudo-Rāzī's *On Alums and Salts* described ammoniac salt as coming "from the liquid and dry excrements from animals" and said that it "gathers especially in the places of dry baths (saunas) as a result of [human] sweat, and in the bottom of a cauldron and in that channel through which fire is sent under cauldrons" (Arbuthnott forthcoming: section 16).

The sulfurous smells of metal workshops where different metals were smelted was akin to the sulfurous smells emanating from volcanoes, and even from Hell to the extent that it had a physical location, as we see vividly portrayed in Michael Scot's discussion of sulfur in his *Liber particularis* (early thirteenth century):

> Each kind of sulfur has certain virtues of great strength, as in alchemy, for changing metals ... and when [sulfur] burns it renders the air fetid (*fetulentus*) ... Many believe that Hell (*infernus*) or the mouth of Hell is [in sulfurous places] because of such great flames, the smell of sulfur and the lamentable cries of the people which are continuously heard by those who live close to them, as well as the sounds of different factories (*officinae*).
>
> (Voskoboynikov 2019: 264–5)

THE MORAL DANGERS OF ALCHEMY TO THE COMMUNITY AND THE INDIVIDUAL

Iohannes Buridan considered the practice of alchemy in a monastic community to be unacceptable precisely because of its practical nature (it did not involve intellectual speculation):

> It is not good for the whole community that each person should be involved in all the practical sciences. For, if all (the monks) were practitioners and none of them theoreticians (*speculativi*) the community would not have such glory and attractiveness as it would if some were theoreticians and others practitioners ... If they were all alchemists and expert in many other practical sciences, it would go badly for the community.
>
> (Patar 1991: 176)

Here alchemy was regarded as the archetypal practical art (see Chapter 4). A few years later, Nicolas Oresme wrote virtually the same thing, but added to alchemy the art of casting spells (*ars incantatoria*) and the foreknowledge of the future (*praescientia futurorum*), "assuming that they are possible" (Patar 1995: 103, line 68). The danger of the practice of alchemy to the individual was put simply by St Bonaventura in the late thirteenth century, when he claimed that "many people in their superstition are made foolish. I have heard about someone who, wishing to become an expert in alchemy, became foolish" (Bougerol 1993: I:134). But a more circumspect account is told by his contemporary, Gerhardus de Fracheto, who mentions

> a certain elderly person in the Order, being literate and eloquent and very pleasing to people in authority, through his love of a certain brother in the flesh arrived at such a miserable state that he left the Order, and, being intent on alchemy, in order to enrich this brother, went to Sardinia, because he had heard that there were mines there which were useful for alchemy, and because he could more safely lay hidden; for the Brothers did not have a residence on that island. When he had rashly spent a year and more in this kind of deceit, being ill and near to death, when he could find no Brother (of the Order), he said to two clerks who were vagabonds with him: "Behold, my dear friends, I die outside the sacred order, which as a wretched carnal man I left. But I have the habit of my order in my bag, which, I beg you to put on me as quickly as possible, so that at least I should be buried in it." But when they tried to do this, suddenly so many lice swarmed from his body that the clerks, terrified and covered with those vermin, fled, and were not able to completely cover his cadaver with earth because of the number of lice.
>
> (Reichert 1896: 290–1)

But formal legal processes against practicing alchemy were not, it seems, very common. The most detailed prescription against alchemy is the one mentioned by Nicholas Eymerich as a decretal of Pope John XXII, "*Spondent quas non exhibent*" (ca. 1317; Latin and French translations in Halleux 1979: 124–5, and discussed below), and here the accusation was of making false gold and silver rather than condemning alchemy as such. Instead of being outright condemned, alchemy was subject to debate – especially in regard to the possibility of transmutation (see Chapter 4).

An anonymous continuer of Thomas Aquinas' commentary on Aristotle's *Meteora* admits that there is a "true" alchemy, but it is very difficult to practice. They "sometimes produce a true generation of metals, sometimes from the aforementioned sulfur and mercury without the generation of an exhalation, but at other times by making such a vaporous exhalation 'sweat' from certain bodies, through the application of proportionate heat, which is a natural agent" (*Continuatio Thomae de Aquino Expositionis in Aristotelis libros Meteorologicorum*, 3, 9; Anonymous 1886: cx, col. 2).

Sometimes it was considered that a successful operation resulted from cooperation with a demon or some sacrilegious act. The early picture book about the different manifestations of the Antichrist – *Das Puch von dem Entkrist* – printed around 1470, includes a woodcut of the Antichrist with Satan hovering over his head in the laboratory of his assistant, an alchemist (Ogrinc 1980: 118).

The objections to alchemy are nicely summed up in Sebastian Brant's *Stultifera navis* ("Ship of Fools") published in Basel, which includes a woodcut in its 1498 edition illustrating "Counterfeiters and Deception" (Figure 5.1), which depicts an alchemist at his furnace. Among the accompanying words are:

> The alchemists qualify for this name, since species of things cannot be changed. Whoever therefore believes that any creature can be changed either into something better or into something worse, or transformed into another species or another likeness, unless by the Creator Himself who made all things and through whom all things were made, without doubt is an infidel and worse than a pagan.

FOLIO

De falſariis & fraudulentia.
Alchimia docet fallax: corruptio vini &:
Quas fraudes hominū perfidus orbis agat.
Omnibus in rebus fraudis falſiq;/doliq;:
Cuncta ſcatent vitio: tuta nec vlla fides.

De falſa-
riis rerum oim

Valeant alchimi ſtę quoniaꝫ ſpēs rerū tranſmutari non pñt. Quiſqs ergo credit poſſe fieri aliq̄ creatu- rā: aut in melius: aut in deteri⁹ im mutari/aut tranſ formari in aliam ſpeciē vel in aliaꝫ ſilitudinē niſi ab ipſo creatore qui oia fecit & p quē oia facta ſunt: p cul dubio infidel' eſt & pagano de- terior.

Cōtra Alchi miſtas.vide tex.in.c.epi circa fi.xxvi q.v.

Iam niſi centenas tribuant in carmine voces
Caſtalię: totidem linguas: vireſqꝫ potentes
Dictandi: nequeam falſos componere ſtultos.
Copia magna quidem eſt illorū: magna caterua

FIGURE 5.1 The alchemist as an illustration of "Counterfeiters and Deception." Sebastian Brant, *Stultifera navis*, Basel, 1498, fol. 115v. © Bibliothèque nationale de France.

ALCHEMY, THE STATE, AND RELIGIOUS INSTITUTIONS

From the very beginning, Arabic alchemy was associated with royalty. The Umayyad prince Khālid b. Yazīd (d. 708) was alleged to be the earliest author of an alchemical work, having learned alchemy from his teacher, Maryānūs (hence the *Liber Morieni* is couched as a dialogue between Maryānūs and Khālid). The ambassador of the Abbasid Caliph al-Manṣūr (r. 754–75), ʻUmara ibn Hamza, is said to have paid a visit to the Byzantine emperor, during which the emperor impressed the ambassador with an alchemical display (Principe 2013: 32–3). Of greater veracity is the fact that, in early-tenth-century al-Andalus, Maslama b. Qāsim al-Qurṭubī was the tutor of ʻAbd Allāh, the son of the Andalusi Caliph ʻAbd al-Raḥmān III (caliph 912–961), and wrote a masterful alchemical work. Thereafter there are many anecdotes about the relations of kings and nobles with alchemists. Edward III, king of England (r. 1327–1377), had a strong interest in alchemy and always had alchemists with him in his court (Ogrinc 1980: 118), partly because he needed money for the Hundred Years War. King John I of Aragon (r. 1387–1396) commissioned a treatise on the search for the philosophers' stone from a certain Jaime Lustrach, and John's successor, King Martin I of Aragon (r. 1396–1410), insisted that it should be completed (Luanco 1980: 125–6; Ryan 2011: 154–5). The French king Charles V's (r. 1364–1380) physician and astrologer, Thomas de Pizan (father of Christine de Pizan), was known for making elixirs from gold and mercury, and the English king Henry VI in the mid-fifteenth century encouraged priests to change (*transubstantiare*) base metal into gold, since they were used to changing bread and wine into the body and blood of Christ (Ogrinc 1980: 119). Finally, the Hungarian kings Vladislav II Jagello (r. 1490–1516) and Louis II (r. 1516–1526) employed a full-time alchemist, Nicolaus Melchior De Sibiu (Cibinensis), who wrote a *Processus sub forma missae* for King Vladislav (Ogrinc 1980: 123; Kiss et al. 2006).[2]

Although alchemists aimed to prolong life and devise other health-giving medicines, their main contact with rulers was in connection with producing coins of precious metals, and the main issues in which they were involved were whether the gold was pure gold or counterfeit. If the coins were acceptable, there was a danger of destabilizing the economy by producing too many of them (Principe 2013: 61–2). If they were not, then the alchemist was in danger of being imprisoned or worse by the ruler. In 1466 or 1488, alchemy was prohibited by the Signoria of Venice in order to check inflation, and an ordinance was drawn up in Nürnberg to the same effect in 1493 (Ogrinc 1980: 123). The prevalence of the attempt to make gold and silver coinage led to the importance of refining the chemist's balance so that it could tell the difference between real and counterfeit gold from subtle differences in weight (Jennemann 1997; Newman 2000).

ALCHEMY AND THE CHURCH

Antoine Calvet

Inasmuch as it was considered a profane art, alchemy was of no concern to the Church (Calvet 2018a). For some scholars, like Vincent of Beauvais and Thomas Aquinas, it was a mechanical art subordinated to physics; for Albert the Great, it stemmed from the study of metals and was accordingly worth studying in order to acquire knowledge of them, without, however, making of it a science in the Aristotelian sense of the term. It was Roger Bacon who first raised alchemy to the rank of an experimental science and a criterion for all others. Among the early Franciscans like, for instance, Bonaventura of Iseo (d. ca. 1280), alchemy took on a Christian bent with its moral intent and its use of Biblical citations; this tendency would intensify until, in the fourteenth century, the literature consisted of texts heavily mixed with religious content.

The proscription of alchemy in religious orders

The Franciscan order counted among its ranks a significant number of alchemists, from Brother Elias (d. 1253), Vicar General of the Order of Friars Minor and successor to its founder St Francis, to the anonymous brethren who may have partly authored the large alchemical corpus attributed to Ramon Llull, a Franciscan tertiary, and including the famous and controversial Jean de Roquetaillade. Alchemical texts are wrongly attributed to Roger Bacon, to Duns Scotus, to Raymond Geoffroy (Raimundus Gaufredus), to William of Ockham, and to Brother Elias, on the grounds that he had introduced the practice of alchemy to the Franciscan Order. In the mid-thirteenth century, Brother Salimbene di Adam from Parma gave the following testimony in his *Chronicle* (Calvet 2018a: 127):

> The eleventh fault of Brother Elias: he fell into dishonour by meddling in alchemy. It is beyond doubt that whenever he heard of brethren who had acquired some knowledge of alchemy in the profane world, either theoretical or practical, he had them summoned to keep them near in Gregory's palace ... There were in this palace several apartments and rooms in which Elias kept these brethren and many others, which was practically equivalent to *consulting a pythoness* [1 Sam. 28:7].

Salimbene's conclusion thus establishes a link between alchemy and the divination that had already been condemned by the church. From 1272 to 1310, many chapters, general and provincial, banned the practice of alchemy. Franciscan rules were very strict at the time, forbidding the possession, reading, copying for oneself or others, handing over, or accepting of books on alchemy and on magic. But these measures did not prove successful, since the Noble Art involved a manual and intellectual labor that was highly valued by

St Francis' rule, on condition that it was not practiced for lucrative purposes. Alchemy must have been perceived by the brethren as a "way of producing non-pecuniary riches" (Pereira 2010: 118–19), but to no avail; the primary vocation of Franciscans is for saving souls, preaching, and teaching the truth of the Church, not for indulging in alchemical experiments. The opponents of alchemy within the order were perhaps as numerous as those who cultivated the Art, both among the Conventuals and among the Spirituals or Fraticelli, like Angelo Clareno. Under such circumstances, it is possible that theoretical alchemy, or at least its rudiments, were taught in Franciscan schools through the commentary to Aristotle's *Meteorologica IV* (whose attribution to Aristotle is now disputed). Traces of this teaching, for example, have been preserved in a commentary to *Meteorologica IV* by Géraud du Pescher, lector of the Toulouse *studium generale* (university) in the years 1332–1333 and teacher of Jean de Roquetaillade.

The Dominicans did not condemn alchemy in such a specific way. There is no doubt that the research on minerals by Albert the Great – his example and his teaching – protected the alchemists within the Order. This is how the *Semita recta*, an alchemical treatise attributed to him, remained in the catalogs of his works until the late fourteenth century, before being struck out in the following century. In spite of this, the Preachers tried to ban alchemy in several of their chapters, just as the Friars Minor did. Rather timid initially, the condemnations of alchemists grew in strength in the same measure as they proved ineffective. In 1273, the General Chapter of the Dominican Order, convened at Pest, forbade any brother to study or teach alchemy, or to possess books on alchemy. In case any books were found, they had to be handed immediately to the prior. The chapter of Bordeaux (1287) even threatened those guilty with imprisonment. Noting that in spite of previous bans the brethren persisted in the practice of alchemy, the chapter of Metz (1312) fulminated with threats of excommunication, and this was repeated in Barcelona in 1323.

In the monasteries (Benedictines, Cistercians, etc.), it was only the Cistercians who made provisions against alchemy. In 1317, the General Chapter decried those who were suspected of cultivating this "false art of alchemy," threatening them with excommunication. In contrast to this attitude, the Order of Saint Benedict and the Carmelites (a mendicant order) did not issue any such decree, even though they counted a number of alchemist brethren among their ranks, like Fr. Dominic of the monastery of Saint Proculus in Bologna, or like the Carmelite Guillaume Sedacer.

From these condemnations we can draw the following conclusion: they tended to originate in the mendicant orders. The Franciscan measures are the most draconian and detailed, evincing a strong antialchemical sentiment within the Order, even as the influence of alchemy among the brethren kept growing. From the time of Salimbene, they associated alchemy with magic, just as the Dominicans linked alchemy with magic on the occasion of the provincial chapter of Narbonne (1272). The impact of these interdictions is hard to measure. Their recurrence gives the impression that the brethren continued their researches and pursued their experiments without worrying too much about reprisals. Furthermore, the jurisdiction of these conventual condemnations is limited to their particular spheres; the Holy See seemed not to pay attention to the phenomenon of alchemy. It would not be until the papacy of John XXII (r. 1316–1334), who was obsessed by the problematic occult sciences, that a decretal concerning alchemy would be published. This decretal, "*Spondent quas non exhibent divitias pauperes Alchimistæ*" ("The poor alchemists promise riches that they cannot deliver"), is found undated among the collection of *Extravagantes communes* attributed to John XXII (Halleux 1979: 124–5). It was probably issued in 1317, but it remained completely unnoticed and unmentioned until it came to the attention of the inquisitor Nicholas Eymerich. In this document, Pope John XXII only targeted the alchemists in an indirect way. Those pitiable alchemists were after something too great for them: "the truth they seek is beyond their reach." The pope did not in any way contest the theory of alchemy; he only noted that its pursuit was futile. What he intended to prove with his text was that the art only succeeds in making false gold and silver by means of a sophistical transmutation. Therefore, those who gave any support to this work were destined to failure, and those who used currency minted with alchemical gold and silver were liable to prosecution just as those who produced counterfeit currency. It seems then that the main objection was not to alchemy and the alchemists, but to forgery and counterfeiters. Another passage from the *Spondent* makes this clear: "these [alchemists] should be forced as a punishment to contribute publicly to the poor, true gold or silver of the same amount as the alchemical product." They were regarded as having transgressed the law.[3]

According to the *Directorium Inquisitorum* (1376) by Nicholas Eymerich, alchemy was not a transgression in itself; he seemed to even ignore the decretal by John XXII. But twenty years later, in the course of some correspondence, Eymerich sought to expand on the question of alchemy. In 1396, replying to his friend the canonist Bernat Estruç, abbot of Santa Maria de Roses, who had been troubled by an alchemist, Eymerich composed a little treatise, the *Contra alchimistas*. In this text, the inquisitor compiles most of the criticisms directed at the alchemists, and he finishes his argument with the *Spondent quas non exhibent* of John XXII. The commerce of alchemical gold gives rise to counterfeiting. This gold does not stand the tests for authenticity. He ascribes to demons the power not to create gold, but to transport gold and silver from place to place, "because they know where the treasures are to be found." The *Contra alchimistas* of Nicholas Eymerich had a very limited diffusion during the Middle Ages, and it would be quite an exaggeration to call it the handbook of inquisitors, even more so when we consider that jurists generally declined to condemn the alchemical art whenever they were asked to decide on its legality or illegality.

In the second half of the thirteenth century, in the *Pearl of the Decretum* (*Margarita decreti*), a gloss added to Gratian's *Decretum*, Martin of Poland stated the four positions regarding alchemy that could be drawn upon to seek an indictment: "the first two are those of the accusation, and the other two are those of the defence" (Matton 2009: 12–13). After examining the four arguments in the *Pearl*, Oldrado da Ponte came to the following conclusion that alchemists do not sin (Calvet 2018a: 127):

> If, by means of that virtue which is in herbs, stones or elements, they want to make silver out of tin, since being as they are almost of the same principle and similar to each other, the passage is easier between things that share a commonality.

Thus Oldrado acknowledged as authoritative the basis itself of alchemical theory (i.e. that metals share a common origin). He reinforced his verdict with the Augustinian doctrine of seminal reasons. All doctors of the two laws followed Oldrado, and Giovanni d'Andrea (ca. 1275–1348) contributed an additional historical argument: the transmutations allegedly effected at the Roman Curia by Master Arnald of Villanova, physician of Boniface VIII. All of them or most of them agreed with each other.

The theoretical and practical legitimacy of alchemy was thus accepted throughout the Middle Ages; the few admonitions coming from the Holy See and the Inquisition focus on the counterfeiting of currency potentially endorsed by alchemy, and on the very occasional occurrence of nigromancy (invocation of demons). The licit character of the transmutative art encouraged the alchemists to continue to write works of increasing complexity, using concepts, methods, and ideas borrowed from philosophy, and even from medicine, exegesis, and theology. This notwithstanding, any brief perusal of the texts – mostly those originating among the Franciscans, the pseudo-Llullians and pseudo-Arnaldians, or those of Jean de Roquetaillade – suffices to reveal the dim echoes of the condemnations issued by the General Chapters.

EFFECTS OF CHEMISTRY ON THE PHYSICAL ENVIRONMENT

Justine Bayley

As well as considering the social and intellectual environment in which chemistry and alchemy took place, it is also necessary to look at its impact on the physical environment in which people lived and worked in medieval times. Today, we are very conscious of pollution of all sorts and have many concerns about the effects it has on us and our surroundings. In earlier times, pollution was perceived as less of an issue for various reasons, which are explored below.

So what is meant here by pollution? Most but not all pollutants are human-made, but they were often ignored or deliberately overlooked as they were only a cause for concern if the result of the pollution was harmful. Thus in a medieval context they are perhaps best defined as something that is unwanted or dangerous or has negative impacts on people or society.

Different types of pollution can be detected by the five senses: sight, hearing, smell, taste, and touch. Because the medieval population was far lower than today and most towns and cities were quite small, there was generally more room for people to keep away from pollution hot spots. Thus even when the noxious effects of pollutants were recognized, they were often ignored, as physical separation was sufficient to protect the majority of the population.

Chemical processes like those described in Chapter 6 made materials or things that were the desired products. However, they also produced by-products and waste materials that, if they could not be reused in some way, were discarded and therefore potentially polluted the workshop and its surroundings. The consequences of this disposal could be visible: for example, iron smelting slags were usually just dumped in large heaps as they were of no further use (until new processes to rework them and extract more of the metal were developed in subsequent centuries). They would not have been considered pollution as, other than blocking the view, they had no negative effects.

Nonferrous metals were far less benign, as they have negative impacts on biological matter such as plants and animals. Lead in particular is toxic and is a cumulative poison in animals, including humans. Early lead smelting sites can still be identified as the vegetation downwind from them is sparser and less species-rich than in the general area, as few plants can tolerate even low lead concentrations (Buchanan 1992). Mapping soil lead contents across areas where lead ores were mined has helped identify many sites where lead was smelted (e.g. Wild and Eastwood 1992). The craftsmen smelting lead ores, cupelling the metal to extract silver, or refining silver by cupellation (see Chapter 6) would all have ingested significant quantities of lead compounds, but the link to early deaths would have been masked by generally low life expectancy. Occupational health problems like these were not at the time linked to pollution but were accepted as facts of life; you might otherwise have been trampled to death by a cow or killed by rotten food or an infected wound. In general, medieval people appear to have accepted their fate, believing it to be the unalterable will of God.

Some chemical processes dealing with organic materials, such as tanning leather, retting flax, or dyeing cloth, were smelly, as they involved controlled rotting or fermentation. For example, the first process in tanning hides was to wash them to remove blood and dung, often in a local stream or river. This polluted the watercourse and could lead to complaints from other users like brewers (e.g. Page and Round 1907: 459). Subsequent dehairing of the hides involved leaving them somewhere warm for the hair roots to rot or applying urine. Later steps in tanning involved soaking with bird droppings, dog dung, fermented barley or rye, stale beer, or more urine; all these materials made tanneries unpleasant neighbors (Cherry 1991: 296)! Although smells of all sorts were far more prevalent in medieval times, these unpleasant processes that relied on natural chemical changes to the materials being prepared were normally sited on the edges of settlements on the downwind side so the unpleasant smells were blown away. Despite this general zoning of chemical industries to minimize the nuisance they caused, there is some evidence of them in the centers of settlements. An example is the discovery of woad seeds and leaves on the Coppergate site in Viking York (Kenward and Hall 1995: 714), where presumably woad was used to dye cloth.

Many chemical processes required the use of fire to create high temperatures. When the main fuel was wood (or charcoal), smoke from fires created little pollution. Toward the end of the medieval period coal was increasingly used, as in many areas there was a shortage of wood, and the smoke from coal fires was an atmospheric pollutant – though this only became a major nuisance in towns from the nineteenth century onwards.

Fire was often considered a hazard when buildings were mainly timber built. However, fires used by craftsmen in their industrial processes were normally carefully controlled within a furnace or hearth so were not hazardous or polluting. Most accidental conflagrations were domestic fires that got out of control.

CONCLUSIONS CONCERNING ALCHEMY AND SOCIETY

Alchemy provided a powerful metaphor for all kinds of miraculous changes. Chaucer used a "richly complex alchemical metaphor" in describing Troilus' love in the fourth book of his Troilus and Criseyde (Gabrovsky 2015: 23, 129–98). John Wycliffe in the mid-fourteenth century could compare Christ with the elixir:

> Christ takes away the sins of the world unequivocally, ... since he is the only immaculate principle by which the whole race of those who should be saved is brought to life, just as unclean metal is purged, according to the alchemical Work.
>
> (Iohannes Wyclif, Sermones I, sermo 8; Loserth 1887–1890)

The alchemists also made the comparison: Pseudo-Arnald of Villanova in his *De secretis naturae* regarded the life of Christ as a metaphor for alchemical sublimation (Grabovsky 2015: 72–3), and Alexander of Hales in the early thirteenth century could use alchemy as a metaphor for the process of transubstantiation:

> An example can be taken from the alchemical process: there is a certain body that takes the place of matter, another, that of form, and when spirit dominates over matter, nobler bodies are produced from less noble bodies, and this process is called "sublimation."
> (Alexander of Hales, *Quaestiones disputatae "antequam esset frater"*, q. 64; Alexander of Hales 1960)

Alchemy had come a long way since the initial days in the mid-twelfth century when the meaning of the word was barely understood.

CHAPTER SIX

Trade and Industry: *Medieval Craftsmanship and Technology Transfer*

JUSTINE BAYLEY WITH SPIKE BUCKLOW

INTRODUCTION

For Europe and the Near East there are effectively three manufacturing traditions in the medieval period. The first is the continuation of Classical (Roman and Greek) technologies and is exemplified by the Byzantine Empire. In northern Europe some of these technologies continued, but others disappeared for hundreds of years until they were reintroduced from the Mediterranean world. The Migration Period brought new peoples and their technologies from the east to Europe. At around the same time the rise of Islam in the Middle East brought about a new way of looking at the world and encouraged the development of previously unknown technologies, some of them originating in India or China, areas with which the Arabs had traded for centuries. By the end of the eighth century the Islamic lands covered a vast region, from Spain in the west to the Indus valley in the east, and included all of North Africa and the Near and Middle East. All these areas were in close contact, so unsurprisingly there were many trade links, and the industries of the area shared many of their technologies and even sources of raw materials. By the end of the Middle Ages Western Europe had become a center for industrial and economic innovation. This is exemplified by the banking systems that developed, facilitating trade, and in the rediscovery or adoption and development of technologies of the Classical and Islamic worlds.

In this chapter industry is defined as the production of materials or objects, either for use locally or to be traded elsewhere. Industries transform raw materials into products using a range of processes. The processes considered here are those that produce changes in the chemical composition (in modern terms) of the raw material; some are identified in Chapter 2. If the product of a process was not a finished commodity or object, it was either the raw material for another process or industry or a by-product (waste).

In the medieval period much industry was small-scale work by individuals or small groups of craftsmen rather than the large-scale manufacturing that we associate with the post-medieval period, and especially the Industrial Revolution of the later eighteenth and nineteenth centuries. Some chemical processes were small-scale, widespread craft activities, often practiced on a domestic scale. Others may have been equally small in scale but were centralized at a limited number of locations; they can be considered specialized or industrialized and often required particular skills or knowledge. In some cases, especially in earlier times, a single craftsman would carry out all the necessary processes, but at other times a number of different craftsmen each carried out a single stage in the conversion of raw materials to the finished product.

In discussing medieval industries a number of other, general points also have to be considered. In this period the craftsmen had an empirical understanding of specific processes, but not at a theoretical or generalized level. Their experimentation was limited, which led to continuity of practice; once an acceptable method (or combination of processes) had been devised there was little incentive to change it. This led to conservatism; although designs of artifacts changed continuously, the processes by which they were made stayed much the same for centuries or even millennia.

Many chemical changes require high temperatures, and achieving these depended on the availability of fuels. In this period fuel was mainly organic (wood, charcoal, peat, dung), though inorganic fuels such as coal or oil had begun to be used in some areas. By the end of the medieval period, if not earlier, woodland was being coppiced to provide a regular supply of fuel. Its availability often determined the location of industries; it could be more economic to transport raw materials to the fuel than vice versa.

Some chemical raw materials occur naturally and just have to be collected and, if necessary, purified. Others have to be manufactured by extracting, combining, or reacting together some substances that occur naturally themselves. Many raw materials were widespread, so industries usually used those that could be sourced locally. This gave rise to variability in the products manufactured because of the varied composition of the raw materials. Domestic pottery is a good example since the iron content of the clay, combined with the way the pots were fired, dictated whether they turned out white, buff, red, or gray. Other materials were only found in a very limited range of locations, and these were traded to where they were needed from very early on. Examples include gemstones, precious metals, and pigments such as the cobalt minerals used to color glass and glazes blue.

MEDIEVAL CHEMICAL TECHNOLOGIES

Almost all the chemical industries we know from the medieval period were also well known and widely practiced in the preceding centuries. The chemical processes remained the same though their end products can be dated, but only because of changes in their designs and styles of decoration. The ultimate products of medieval industries were artifacts that were traded, sold, bought, and then used by people in their everyday lives. Examples are metal tools and jewelry, glass and ceramic vessels, loaves of bread, and leather shoes. Making these artifacts could involve many different chemical and physical changes to the original raw materials. Chemical changes alter the chemical composition of the material, either separating out one component of it or producing a new substance with a different chemical composition from two or more raw materials. For inorganic materials most chemical changes take place at high temperatures; examples are given in Table 6.1 and described in more detail below.

TABLE 6.1 Some chemical processes and examples of their use.

Industrial process	Explanation
Smelting	Extracting metals from ores
Refining	Removing some impurities (e.g. in metal refining)
Alloying	Melting metals together to give a product (alloy) with different properties
Cementation	Adding an element to a metal in the solid state (e.g. carbon to iron to make steel [carburization]; zinc to copper to make brass)
Lime burning	Turning limestone into lime mortar
Firing ceramics	Turning clay into bricks, tiles, or pottery; adding a glaze (glassy layer) to fired ceramics
Making glass	Reacting silica with alkalis and/or lead oxide, sometimes adding colorants, decolorants, or opacifiers
Making pigments	E.g. Egyptian blue; many others occur naturally
Dyeing	Coloring organic materials
Fermentation	E.g. making alcohol
Distillation	E.g. separating liquids such as perfumes, alcohol, or strong acids

In addition to these chemical changes there were a whole range of other transformations that were also used by craftsmen – those we now describe as physical changes. The Introduction in this volume has already explained this modern distinction between chemical and physical changes and noted its irrelevance for the medieval period, from the point of view of medieval scholars. The craftsmen often used processes that included both chemical and physical changes. For example, once iron had been reduced from its ores (a chemical change), the metal could be shaped into an object like a knife or a sword by hammering, cutting, and grinding. These processes did not change its chemical composition but did change its shape, and could also change other properties such as its hardness. These physical changes were essential parts of most manufacturing industries but are not considered further here unless they also affect the chemical composition of the material being worked.

Generally, organic materials are not considered in this chapter, as most of the changes made to them were not considered "industrial" processes at the time; they were changes every family would have made for themselves. These materials include foodstuffs that, in this period, were either used as they were harvested or were modified by relatively low-temperature chemical reactions we know as cooking. Plant fibers were widely used: cotton and linen were spun and woven into textiles, and other fibers were used to make such essentials as rope. None of these products required chemical changes to the raw materials so are not considered here.

Fermentation, a group of reactions caused by naturally occurring yeasts, was widely used in the preparation of food and alcoholic drink. Distillation to enhance alcohol levels was a physical rather than chemical change, separating alcohol from the water-based mixtures where it formed. This process, developed initially in the Arab world, was also used to produce perfumes and was later used right across Europe to separate inorganic materials as well (see Chapter 2 and see below in this chapter). Other plant materials naturally contain chemicals we now describe as drugs, and these were normally extracted by crushing or soaking the plant in water that might be gently heated. These processes were the province of the herbalist and, as they do not result in chemical changes but just the separation of the active ingredient, they are also not considered here.

Of the useful organic materials, wool was spun and woven into textiles that were then fulled and dyed to change their properties and appearance. Dyes were also used to color plant-based textiles (see below). Some cloths, such as silks and English worsteds, were considered luxury goods and were traded widely across Europe in the High Middle Ages. Other animal products were modified chemically. If animal skins were dried they became stiff and hard, but if they were tanned by soaking in vats of organic or inorganic chemicals they turned into leather – which usually remained pliable and had many different uses. Other animal products such as horn, bone, ivory, and antler were not chemically modified but physically shaped into useful objects, some of which were widely traded.

The processes used

As Table 6.1 shows, many different chemical processes were used; some of them were specific to a particular industry, but others were applied to a range of different raw materials. Each process converted a raw material into a product. This product was either ready for use or may have been the starting point for a subsequent process. For example, metal production and the subsequent processes used to shape the metal into objects and decorate them represent a sequence of changes – both chemical and physical – that are all necessary steps in the manufacture of metal artifacts. This sequence of processes is often described as a *chaîne opératoire* and, though all of them may have been carried out by a single craftsman, there is potential for workers with different skills or experience to carry out individual processes. This potential subdivision of labor can be seen as the ancestor of modern production lines where each step is carried out by a separate craftsman.

The transformation of raw materials into desired products through chemical reactions demonstrates clearly that changes to materials can and do happen; the secret is finding the right combination of materials and processes. In the practical world of industry, experimentation was not common, so once a satisfactory manufacturing process had been developed it often remained in use for hundreds if not thousands of years.

The chemical industries

A wide range of products such as dyes and pigments, paints, metals, glass, ceramics, fuels, soap, fertilizers, spirits and essential oils, foodstuffs, and drugs were made by a variety of processes from combinations of many different raw materials. The processing of organic materials has already been touched on but will not be considered further, as it is not possible to cover all of the classes of products above in adequate depth. The focus will instead be on a few specific products where the raw materials were inorganic. These examples have been chosen as the evidence we have for them suggests that the scale of operation and the perceived specialist nature of the processes involved bring them closer to our modern definition of industries than the smaller-scale and more widespread manufacturing that operated on a domestic scale.

There are multiple sources of evidence for the industries discussed below. Some information comes from archaeological excavations on industrial sites (see Chapter 3), though the number of workshops positively identified is small (Crossley 1967; Bayley 1992; Jouttijärvi et al. 2005; Bourlet and Thomas 2018). From the scientific study of the raw materials and production debris found there and from studies of the products of these industrial sites, whether found there or on related settlements, further details can be deduced. To this material evidence can be added the information contained in surviving medieval documents that shed light on some of the chemical processes carried out. Examples (for the West) of technical treatises for the early medieval period include Theophilus' *De diversis artibus* (Hawthorne and Smith 1979) and the *Mappae clavicula* (Smith and Hawthorne 1974), both mentioned in the Introduction, and by the sixteenth century there are many printed books such as those by Agricola (Hoover and Hoover 1950), Biringuccio (Smith and Gnudi 1966), and Ercker (Sisco and Smith 1951). There are also many Arabic sources, especially for the earlier period, which describe and discuss the products and processes used in the Islamic world that later became commonplace in Europe too. An example is al-Kindi's *On Swords* (Hoyland and Gilmour 2006).

CASE STUDIES

Two major industries – metal- and glass-working – have been chosen for more detailed examination. Although both produced materials and objects that were widely traded and used throughout the medieval period there were great differences between them. Metal technology had evolved to a high level before and during the Classical period so there are relatively few innovations that can be attributed to medieval craftsmen. Metalworkers across the area that had been part of the Roman Empire generally used similar technologies, but some regional specializations developed. On the other hand, the glass that was used in the Roman period had one general chemical composition, but the following millennium saw the introduction of many different glass compositions, each with distinct geographical origins and distributions.

Metal production

Most metals are not found in their native state as they are reactive elements. Instead they are found in the earth's crust as ores, natural occurrences of rock or sediment that contain sufficient quantities of metal compounds (in the form of minerals) to be extracted from the deposit economically. Typical minerals are hematite (iron oxide, Fe_2O_3) and goethite (hydrated iron oxide, FeO[OH]), the copper minerals malachite ($Cu_2CO_3[OH]_2$), chalcopyrite ($CuFeS_2$), and tennantite ($Cu_{12}As_4S_{13}$), galena (lead sulfide, PbS), and cassiterite (tin oxide, SnO_2). In order to extract the metal from the ore, the mineral has to be reduced to the metal, and the metal has to be separated from the rock in which the mineral was found.

The process by which metals are obtained from their ores is called smelting (see Figure 6.1), but in practice this must be preceded by mining and ore beneficiation, where the mined rock is crushed and sorted to select only the mineral-rich parts. The enriched ore may also be roasted to convert the minerals

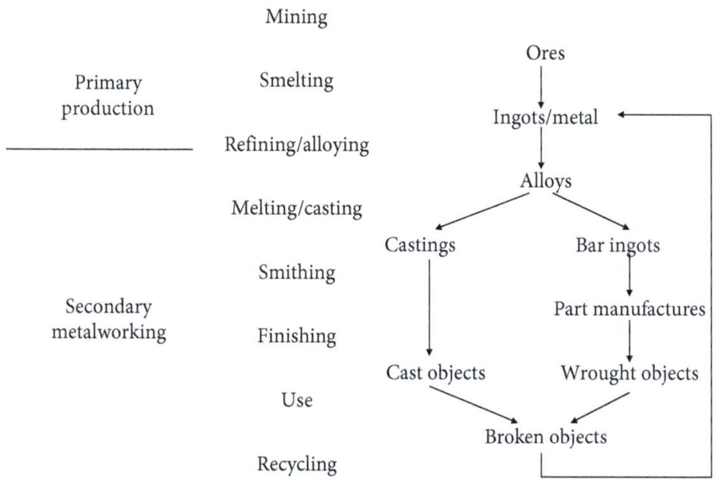

FIGURE 6.1 Metalworking processes. The central column lists various processes that relate to the primary production of metals and alloys (above) and those carried out during secondary working, turning metals into useable objects (below). To the right are listed the products of each process, which are then the raw material for the next process: for example, mining provides ores, which are the raw material for smelting. Diagram by Justine Bayley.

to oxides and drive off any chemically bound water before it is then smelted in a furnace, which was usually made of or lined with clay and was blown with bellows to create a much hotter temperature. The details of the smelting process depend on the metal being extracted and the mineral from which it comes.

Ores and smelting slag are, respectively, the main raw materials and by-products of metallurgy. The morphology, chemistry, and microstructure of slags can yield information about the metals being produced and underlying engineering parameters such as furnace types, operating temperatures and redox conditions, the selection of natural materials, and the efficiency of ancient technologies. It is thus possible to use slag analyses to reconstruct knowledge transmission, technological traditions, and their evolution in time and space in response to different environmental and cultural constraints.

For smelting copper in antiquity there were usually three stages to the process: roasting, smelting, and purification of the metal that was produced. The charcoal fire in the furnace reaches a temperature of the order of 1,000°C and, despite all the air being blown into the furnace, the charcoal burns incompletely to give carbon monoxide (the atmosphere is essentially reducing), which reacts with oxide minerals to reduce them to metals. At this operating temperature the copper melts and collects in the cooler base of the furnace or is run out of it (Tylecote 1986: 16–22). Many copper minerals, however, are sulfides, and roasting burns off some of the sulfur. Smelting then involves a series of reactions between the various minerals present to produce copper (Craddock 1995: 135–7). This metal contained impurities so had to be refined before it was used (Craddock 1995: 202–4). A more efficient smelting process for low-grade sulfide ores involves making matte (copper sulfide, CuS) as an intermediate product of what is a complex multistage process (Craddock 1995: 149–53). This is a later development – quite when it came into general use in Europe is not clear, though the descriptions in sixteenth-century treatises show it was well understood and therefore probably not a recent innovation (Hoover and Hoover 1950; Smith and Gnudi 1966).

Iron smelting is also complex, partly because pure iron metal melts only above 1,500°C, a temperature that could not be reached in early furnaces. Instead the iron oxides were reduced to metallic iron that remained in the solid state and was collected together into what is known as a bloom, while the rock that had contained the iron ore reacted with some of the iron oxides to form a slag that was liquid at around 1,100°C and so could be separated from the bloom. The iron ore and the charcoal were added gradually to the top of the furnace and as they moved down the shaft the temperature rose and both the reduction of the iron ore and the formation of slag began. The iron bloom began to form just below the level of the tuyere (the opening where the air from the bellows entered the furnace) and in medieval bloomery furnaces the slag ran down to the bottom of the furnace and was tapped out. Because some of the iron in the ore was lost into the fayalitic slag (Fe_2SiO_4), it was practical to smelt only iron-rich ores. This bloomery smelting process was the only one known until the later medieval period when the process was modified by adding water-powered bellows, which allowed larger blooms to be produced; the scale of production was increasing. From this innovation a new type of furnace – the blast furnace – evolved.

While blast furnaces were not in use in Britain until the very end of the fifteenth century (Awty 2003), they were in use as early as the thirteenth century in Sweden and Germany (Magnusson 1985; Jockenhövel 2013), and they are said to have been in use in Muslim lands by the tenth century (al-Hassan and Hill 1986: 259). Unlike late bloomery furnaces they always employed water-powered bellows, and they were taller to allow enough time for the chemical reactions to take place as the charge gradually descended. They worked at higher temperatures than bloomeries and the atmosphere in the furnace was more strongly reducing so almost all of the iron in the ore was reduced to metal. The chemistry of the slag therefore changed as, instead of iron, it contained mainly lime, added with the fuel and ore as limestone, giving a calcium silicate slag that melted at a higher temperature than fayalitic slags. The other important change was that as the atmosphere within the furnace was far more reducing, carbon diffused into the metallic iron, reducing its melting point. Thus, both the metal and the slag were liquid and could be tapped out of the blast furnace. The metal was sometimes cast directly into large objects such as cannon but was usually cast into ingots known as pigs.

The iron produced by consolidating a bloom had a very different chemistry from that of a cast pig. Bloomery iron, often known as wrought iron, was usually very pure iron with little if any carbon dissolved in it. It was relatively soft and easy to hammer into shape when red hot, and it could be welded together at bright-red heat. Sometimes low levels of carbon (up to 1 percent or so) had diffused into the iron in the smelting furnace and changed its physical properties, making it harder and better able to keep a sharp edge but more brittle – this iron alloy is known as steel. If the ore from which the iron was made contained phosphorus, this element would be present in the smelted metal. This was an advantage, as phosphoric iron was harder than pure iron, but it also stopped the diffusion of carbon into the iron so it could not be turned into steel. Cast iron, on the other hand, contained up to 5 percent carbon in the form of either graphite (elemental carbon) or cementite (iron carbide, Fe_3C). The presence of these large amounts of carbon makes the metal brittle so it cannot be hammered into shape. Although some iron was remelted and cast into objects such as fire backs, most of the pig iron was decarburized through a process known as fining. Depending on how far this process was taken, the end product was either steel or pure iron, both of which could be hammered into objects.

Across the Middle East steel was also made by melting together cast iron and wrought iron in a lidded crucible, a process that originated in the Indian subcontinent and Central Asia (Craddock 2003). Crucibles dated to the ninth century are known from Merv (Turkmenistan; Feuerbach et al. 2003), and the process was described by al-Bīrūnī in the tenth century (al-Hassan and Hill 1986: 254). This high-quality steel, sometimes called *wootz*, contained 1–2 percent carbon and was valued across the Islamic world. It was widely traded by Persian merchants and used to make sword blades, which were known as Damascus steel, though they were made in other places too. There are written sources that indicate that crucible steel was used from the late tenth century and also worked from at least the twelfth century in al-Andalus (Spain); there is one late thirteenth-century text that mentions making crucible steel in Ceuta but as yet no archaeological evidence of this (Dinnetz 2001).

Lead metal has a much lower melting point (327°C) and is easily smelted from its principal ore, galena (PbS), in a low-temperature wood fire (Tylecote 1986: 54–8; Craddock 1995: 205–11). Lead was used as a metal in its own right, and was often added to copper alloys that were to be used for casting. However, lead ores were also the principal sources for silver, which was extracted from newly smelted lead through a process known as cupellation (Bayley and Eckstein 1997; see also Chapter 2). The lead metal was melted on an open hearth lined with an absorbent material. The conditions were oxidizing, so the lead was converted to lead oxide (litharge, PbO) that was either raked off the surface of the pool of molten metal or was absorbed by the hearth lining. Silver oxidizes far less readily than lead, so it remained unaltered on the surface of the hearth at the end of the process. A potential problem was that lead oxides react with clay (the normal material used to line hearths) to form a glass, which would seal the surface and stop any absorption of the litharge. Cupellation hearths were therefore lined with vegetable ash, which was sufficiently porous to absorb the litharge but did not react chemically with it. The hearth lining could be resmelted to reclaim the lead metal.

Cupellation was also used to refine silver that was being recycled. The debased silver was melted with an excess of lead under oxidizing conditions and the lead oxidized to litharge, which acts in two ways: it reacts with the base metals mixed with the silver, oxidizing them, and it also forms fusible compounds with these metal oxides, which do not inhibit the oxidation of the remaining base metals (Percy 1870: 177–8). As noted above, the silver does not react and the litharge and base metal oxides are absorbed by the hearth lining, which is either bone ash or calcareous clay (Bayley and Eckstein 2006). Exactly the same process was also used when assaying (testing the purity of) silver, but instead of using a lined hearth the reaction took place in a small cupel made of bone ash. In this case a small sample of metal was weighed, cupelled, and then weighed again; the difference in weight measured the purity of the metal (Bayley 2008). There are no surviving bone ash cupels from early medieval times, but shallow ceramic dishes of a similar size appear to have been used, though, as noted for large hearths, the lead-rich glassy surface could trap some of the silver (Bayley 2008) (Figure 6.2). No archaeological evidence that gold was cupelled has been found, but there are a few early medieval examples of small, highly refractory "ceramics" where gold appears to have been melted with some sort of lead-free flux to remove impurities (Bayley 1992: 750–1, plate LVa; Lamm 2008: fig. 26; Rehren and Nixon 2014).

Lead was also used in the process known as lead soaking or liquation, which separates silver alloyed with copper. The argentiferous copper was melted with excess lead, the silver preferentially dissolved in the lead, and the metal was cast into ingots. Lead and copper are almost completely insoluble so they are present as two separate phases. When the ingots were heated gently the copper remained solid, but because of its much lower melting point the lead liquefied and trickled out, taking the silver with it. The lead could be cupelled to regain the silver and the "sponge" of copper melted to make useable metal (L'Hertier and Téreygeol 2010).

Although cupellation separates base from precious metals, it cannot separate gold from silver. Parting was the name given to a group of processes used to separate gold and silver from at least the mid-first millennium BCE (Ramage and Craddock 2000). Until the medieval period the process was a solid-state (cementation) reaction known as salt parting. All the archaeological evidence

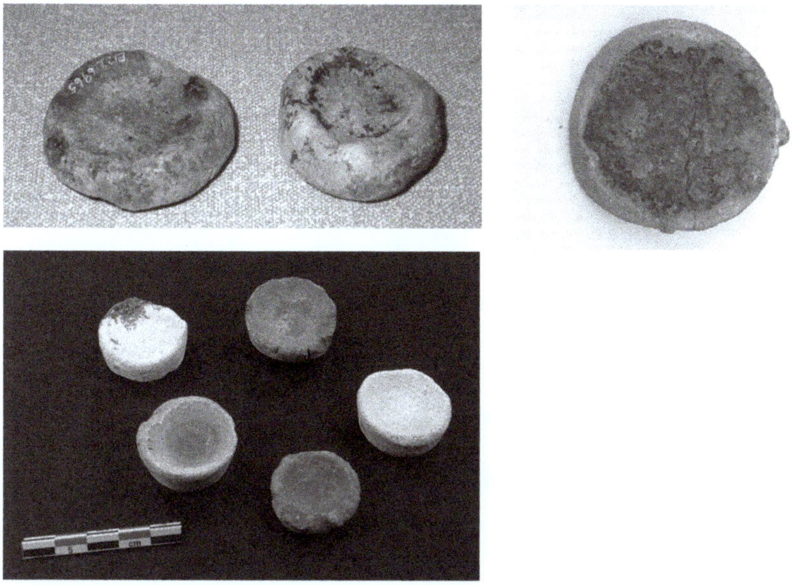

FIGURE 6.2 Cupels from the British Isles. Those at the top are early medieval ceramics, while those at the bottom are used (dark) and unused (light) sixteenth-century examples made of bone ash. The scales are in millimeters/centimeters. Top © Justine Bayley; bottom © Nicolas Thomas.

indicates that the active ingredient was common salt (NaCl), though surviving texts indicate other salts could be used (Hoover and Hoover 1950: 453–7). The salt was mixed with an inert material such as powdered ceramic to make a "cement" that interleaved thin sheets of the precious metal in pots known as parting vessels that were then sealed and heated below the melting point of the metal (Bayley 2008).[1] These vessels functioned in the same way as those used for cementation (cf. Figure 2.2), though in parting the objective was to separate from rather than to add to the metal. The salt (NaCl) decomposed at high temperatures in the presence of silica and alumina (the ceramic) to give hydrochloric acid and, probably, free chlorine, which reacted with the silver in the metal, forming silver chloride (AgCl). Moisture, presumably from the combustion of the fuel, is another unspecified but vital reactant. Silver chloride is volatile and so was removed from the metal, leaving behind the purified gold. The lid was an essential part of the parting vessel as it prevented the loss of volatile reactants, allowing the silver chloride to be absorbed by the cement and, to a lesser extent, by the walls of the parting vessel. The silver could be recovered by cupelling the cement afterwards.

In late medieval times distillation was introduced into Western Europe (Taylor and Singer 1956; Forbes 1970).[2] In Britain there are documentary references from the fourteenth century and archaeological finds of definite

FIGURE 6.3 A sixteenth-century assay laboratory with furnaces for distilling strong mineral acids (center foreground). Note at the back a pair of crucible furnaces and, top left, an assaying furnace. Lazarus Ercker, *Fleta minor. The Laws of Art and Nature*, 1683, book 1, p. 2. Source: Wellcome Collection.

distillation vessels from the fifteenth century or later (Tyson 2000: 168ff), but there is no evidence for the production of strong mineral acids before the sixteenth century (Bayley 2008; contra Greenaway 1972). These were important metallurgically as *aqua valens* (nitric acid, HNO_3) was used to part silver from gold by dissolving the silver while leaving the gold unaltered. Reacting saltpeter (KNO_3) with green vitriol ($FeSO_4$) or alum ($KAl[SO_4]_2$) in a cucurbit produced nitric acid, which could be separated by distillation (Figure 6.3). Archaeological finds of cucurbits are notable for the deposits of red iron oxides inside them. It is these deposits that identify the reaction and hence the product (Bayley and White 2013). Once nitric acid became available the old, salt parting process was superseded and went out of use.

Glassmaking

Glass has been made in the Middle East since the mid-third millennium BCE. These early glasses were rare, were almost universally opaque and strongly colored, and were regarded as artificial gemstones. From the mid-second millennium BCE core-formed glass vessels were made on a significant scale, but once glassblowing was discovered in the coastal region of Syria in the first century BCE, the quantity of glass made increased enormously, and thin-walled, transparent, and colorless glass became common throughout the Roman Empire (Freestone 1991: 38).

Although glass is a solid, the atoms within it are not regularly arranged in a crystalline structure but are randomly distributed like those in a liquid. When it is melted it can be easily shaped, and when it solidifies it can be cut and polished. Glass was usually made by heating silica (SiO_2) with an alkali flux, essentially soda (Na_2CO_3) or potash (K_2CO_3), though they always contained other minor constituents too, at a temperature of around 1,000°C. The alkalis could be naturally occurring minerals or derived from ashes from the burning of plants or wood. Glass made from very pure materials is chemically unstable but the alkali fluxes used in medieval and earlier times had lime present in them, or crushed shells (another source of calcium carbonate, $CaCO_3$) could be present in the silica sand, which stabilized the glass. The batch that was melted thus contained a range of minerals, making the glass much more durable. The glass naturally had a pale, usually greenish hue because it also contained low levels of iron, a contaminant in the raw materials. Sometimes other metal oxides were deliberately added to the batch to give the glass a deep color (see below). Sometimes lead oxide (PbO) was used as the flux instead of or combined with the alkali, producing different types of glass.

Over the millennium or so that comprises the medieval period glasses were made using a wide range of different fluxes. Most recipes are specific to a particular area or period, so if archaeological glass finds are analyzed, their composition can be used to help date them and indicate where they were made. Early glass was made using soda-rich plant ashes, but for most of the first millennium CE Palestine and Egypt supplied the Mediterranean and Europe with virtually all of its glass, made using mineral soda (natron, mainly $Na_2CO_3{\cdot}10H_2O$) from the Wadi Natrun in Egypt. The natron glass was manufactured in very large furnaces in Egypt and the Levant, and the raw glass that they produced was then traded across the Roman Empire to where it was remelted and blown into vessels or shaped into window glass, trinkets, or mosaic tesserae. After the collapse of the Roman Empire in the West, glassmaking and recycling became far less uniform as different areas devised their own strategies. A range of primary glass production compositions have been identified in the Mediterranean and beyond, dating to the late antique and early medieval periods (Schibille et al. 2016). They indicate the use of different raw materials, particularly silica sources, which can be linked to different primary production locations.

In the East, namely in the Byzantine Empire, glass continued to be made in both the Levant and Egypt to the same basic recipes, despite the areas coming under Islamic rule from 640 CE. In the early eighth century there were major political, cultural, and social changes, and minor changes can also be seen in the composition of the glass being made. Initially the glassmaking furnaces moved away from the Levantine coast and the raw materials they used altered slightly. There were then problems with sourcing natron, which drove a technological change to plant ash glass in the later eighth century. In Egypt natron glass continued to be made for another century, but eventually plant ash glass came back into use there too, probably because of a reduction in the quantities of natron available; the changes were long term and gradual, so they were probably related to issues at the natron source (Phelps et al. 2016). Islamic glass weights are one type of object that can be precisely dated to within a few years from the date or name embossed on them. This has enabled the developments of medieval and early Islamic glass compositions to be traced in great detail (Gratuze and Barrandon 1990).

Soda-rich plant ash glasses were made in the Sasanian Empire much earlier than this, but their composition is higher in magnesia (Freestone 2006), and Brill (2005: 75) has concluded that there is a general "overlapping but no close agreement" between the Sasanian and later Islamic glass compositions. Mesopotamian regional production of plant ash glass provided vitreous materials for the ornamentation of the palace-city of Samarra, built in the ninth century CE. The similarities with earlier glasses suggest that the glassmakers of the early Abbasid caliphate continued a centuries-old Sasanian glassmaking tradition to obtain near-colorless glass (Schibille et al. 2018). Despite the large-scale production and interregional trade of Mesopotamian plant ash glasses during the ninth century CE, other plant ash glass found at Samarra originated in the Levant or Egypt. Glass tesserae made of recycled or reused glass were also imported from Europe or the Mediterranean. It is widely assumed that wall mosaics were intrinsically Byzantine, and in the early eighth century the Byzantine emperor allegedly supplied the materials and workforce for the mosaic decoration of the great Umayyad mosques in Damascus and Medina (James 2006). Support for this claim is found in numerous Arab sources,[3] according to which the Byzantine emperor sent money, glass tesserae, and workmen at the request of the Caliph al-Walīd for the decoration of the Great Mosque in Medina (Gibb 1958).

Glass widely used in the late Roman world was still a natron glass, but it contained higher levels of iron, manganese, and titanium, giving it an olive or yellowish-green hue; it is believed to have been made in Egypt (Freestone et al. 2005). A similar glass composition continued to be in widespread use in Northern Europe in the fifth to sixth centuries, suggesting that the glassworkers there continued to access the international trade in raw glass, but from ca. 550 CE this glass was recycled with added "potash-rich" material, designed to increase the dwindling stocks of glass as new supplies from the Mediterranean were no longer available (Freestone et al. 2008: 40–1). Across Europe from the late seventh century there is evidence for an alternative strategy for dealing with the lack of new glass; the relatively high concentrations of colorant elements suggest that Roman glass, including highly colored material, was being recycled. This practice was not sustainable in the longer term, so by the ninth to tenth centuries a new type of glass had come into general use in Europe. The flux was no longer soda-rich but dominated by potash, indicating the use of ash from

FIGURE 6.4 A Bohemian glass furnace of the early fifteenth century CE. Illustration from a manuscript of Sir John Mandeville's travels (BL, MS Add 24189, fol. 16). Source: Alamy.

terrestrial plants or trees. This type of glass predominated in Europe to the end of the medieval period and beyond (Figure 6.4).

As with the earlier soda-based glasses, there were regional variations in composition, and for the later medieval period three types have been distinguished. These are associated with glass production in northwestern France, in the region around the Rhine, and in Central Europe (Adlington et al. 2019). Window glass from New College, Oxford, dating to the late fourteenth century, was from two areas: the colored glass was from the Rhine region, while the white glass is consistent with an origin in northwestern France or England. In a slightly earlier window from York Minster the colored and white glasses were identical and consistent with an origin in northwestern France. A single medieval mirror found in Egypt is of Central European origin, and similar mirror glass is known from Italy; this choice may reflect an advantage of the glass, which is low in iron and hence closer to colorless (Adlington et al. 2019). The finest white (i.e. colorless) glass was the *crystallino* made in Venice in the later Middle Ages. Its raw materials were carefully selected pebbles or sand (the silica) and soda ash imported from the Levant. It was a premium product, and the Venetian authorities attempted to restrict both the onward trade in soda ash and the movement of glassmakers to retain their near monopoly.

As mentioned above, lead oxide was also used as a glassmaking flux, but usually only in specific cases where the properties it gave to the glass were necessary. It is present in enamel (glass applied as decoration to metal objects), in glass "gems" and other trinkets, and in many beads, where it allowed the development of deep opaque colors (see below). Most of these glasses contained lead oxide as well as some alkali. Around the tenth century CE an almost pure lead silicate glass appeared in Northern Europe, with the earliest examples in the Slav lands to the east; these can contain as much as 70 percent lead oxide (Besborodov 1957). There, as in Northern and Western Europe, it was used to make beads and small glass rings that were usually translucent yellow, emerald green, or opaque black (Bayley 2009; Siemianowska et al. 2019). The same sort of glass was occasionally used to make vessels in Western Europe in the fourteenth century (Baumgartner and Krueger 1988: 161–75). There are high-lead glasses too in the Mediterranean; they include several emerald green Islamic vessels, thought to have originated in Syria but found on the early eleventh-century Serçe Limanı shipwreck off Turkey (Brill 2009). Krueger has compiled a longer list of high-lead Islamic glass, many of them green vessels (2014), and equates this material with the *mīnā* mentioned by Rāzī in his early tenth-century *Book of Secrets* (Käs 2010: 2:1071).

COLORING

We may think of medieval manufactures in monochrome, but to the people who made and used them color was an important part of their daily lives; you only have to look at any illuminated manuscript to see the range of strong colors that were regularly used. This polychromy was achieved in a variety of ways. Pigments, both naturally occurring and human-made, were applied to wooden objects, plaster walls, and parchment manuscripts. Organic materials like textiles and leather were dyed, glass had colorants added to the melt or enamels applied to glass vessels after they were formed, and even metal objects were colored by surface plating and enameling.

Pigments

Spike Bucklow

One set of processes that lent itself to a simple means of assessing outcomes was the making of pigments. These were made using a wide variety of methods – by artists, apothecaries, and alchemists – for both their colors and their medicinal properties. Judging a product's efficacy as a medicine depended on the vagaries of human responses so was complicated by factors of use. But the efficacy of an artist's material was plain to see. As colors, the products of different processes could be placed side by side and their relative qualities would have been obvious to all, before being complicated by any further processing or use.

Numerous surviving artists' manuals describe how pigments were made. There is no evidence that artists' recipes were required by practitioners as the guild, apprentice, and journeyman systems formalized the transmission of skills through demonstration, observation, and participation. Some recipes were not written by artists but assembled by doctors or lawyers, suggesting theoretical interest rather than practical necessity. Some are presented with other crafts, like metalwork and glassmaking (Merrifield 1967: 47–70, 82–110). Whatever motivations may have lain behind the creation and collation of recipes, they nonetheless provide data about material transformations that are related to alchemical transformations.

Artists' recipes vary considerably in detail. Comprehensive descriptions of ingredients, equipment, and operations exist for cupellation, cementation, sublimation, etc. (see Chapter 2). On the other hand, simple statements like "this is made by alchemy" are applied to pigments that involve obvious material transformations, like ceruse (synthetic basic lead carbonate), and also to those that appear to be simple natural products, like arzica (a yellow dye made from the plant known as weld). The level of detail in recipes was related less to the complexity of the process and more to the product's prestige, with the purification of lapis lazuli, for example, being very well documented (Thompson 1960: 30, 32, 36–9). Since these processes are very repeatable, the omission of details provides an insight into what authors assumed to be common knowledge.

The contents of artists' written recipes can also be compared with the results of analysis undertaken on the products of artists' practice. Since historic paintings are cultural artifacts that have retained their high value for centuries, they have often been well maintained. They are therefore an invaluable archaeological resource for studying the cultural history of chemistry. This is especially true of paintings' inorganic components. Up to the eighteenth century, the physical evidence offered by paintings shows a very high degree of correlation with the processes that are outlined in contemporary artists' manuals. Together, manuals and paintings provide evidence of specific chemical practices and also evidence of changes in those practices, reflecting cultural developments in chemistry as practiced more broadly in society, including eighteenth-century experimentation and nineteenth-century industrialization.

Dyes

Justine Bayley

Most dyes were derived from plant or animal matter. They were used to color textiles and also other organic materials like leather and wood. In order to make permanently colored textiles it was usually necessary to combine the dye with a mordant (dye fixative). The type of mordant used can affect the shade of the final color and also affects the fastness of the dye. Mordants bind dyes on fabrics by forming a compound with the dye that was able to attach itself to the fabric (or other organic material). The most common mordant used in medieval and earlier times was alum, usually potash alum ($KAl[SO_4]_2{\cdot}12H_2O$) or ammonium alum ($NH_4Al[SO_4]_2{\cdot}12H_2O$), but if the alum was contaminated with copperas (ferrous sulfate, $FeSO_4$) this darkened and dulled the dye colors. Copperas was also used to deliberately blacken leather and as a constituent of ink, in combination with extracts of oak galls (tannins). A summary of the dyes commonly used in the medieval period is provided by Taylor and Singer (1956: 364–9).

Coloring glass and glazes

A very limited range of elements were used both in antiquity and in the medieval period to color glass (and glazes on ceramics). Two main groups can be identified: the first where the coloring element dissolved in the glass, giving a transparent color, and the second where the color was due to fine crystals dispersed through the glass. Table 6.2 summarizes the colorants used.

The color produced by a dissolved colorant is influenced by a number of factors, among them the composition of the base glass and the concentration and oxidation state of the coloring compound. Copper (Cu^{2+}) produces a turquoise blue in alkali glasses but a deep green in lead glasses and glazes. Iron produces blue, green, or amber/brown colors, depending on its oxidation state, and in higher concentrations makes the glass look black. Dissolved manganese also acts as a colorant, giving a range of golden brown to pink and purple colors, again depending on its oxidation state. If both iron and manganese are present in the glass there is an interaction (a dynamic redox equilibrium) between them that can produce "colorless" glass (Biek and Bayley 1979: 14–15; Sellner et al. 1979). The majority of medieval window glass, except for reds and blues, is colored by iron and/or manganese.

The alternative coloring mechanism was adding compounds to the base glass that result in the formation of colored crystals. As they are very small and numerous, they cause light traveling through the glass to be scattered so it appears opaque rather than transparent. Examples of crystalline colorants/

TABLE 6.2 Summary of the colorants and opacifiers found in medieval glass and enamel.

Enamel color	**Colorant and opacifier compounds**
Opaque white	Calcium antimonate ($Ca_2Sb_2O_7$) or tin oxide (SnO_2)
Turquoise (pale blue)	Cupric oxide (CuO)
Dark blue	Cobalt oxide (CoO)
Purple and pink	Manganese oxide (MnO)
Black	Iron oxide (FeO)
Opaque red	Cuprous oxide (Cu_2O) or copper (Cu)
Opaque orange	Cuprous oxide (Cu_2O)
Opaque yellow	Lead antimonate ($Pb_2Sb_2O_7$) or lead tin oxide ($PbSnO_3$)
Golden brown	Manganese oxide (MnO) and/or iron oxide (Fe_2O_3)
Green	Iron oxide (FeO) and/or copper oxide (CuO)

opacifiers are white tin oxide (SnO_2), yellow lead tin oxide ($PbSnO_3$), and red cuprous oxide (Cu_2O; Biek and Bayley 1979: 9). A combination of dissolved and crystalline colorants can also be used, such as dissolved copper oxide (turquoise) combined with lead tin oxide (yellow) to produce an opaque green. These opaque glasses were used to make or decorate beads and as enamels on glass vessels and metalwork, while tin oxide was widely used to produce the white glaze applied to maiolica pottery, which then often had designs in other colors painted on before it was fired. In al-Andalus these ceramics were also decorated with metallic lusters (Canby 1997), which were also used, alongside enamels, to decorate glass vessels such as mosque lamps (e.g. Baumgartner and Krueger 1988: 119–25). Gilded and enameled glass was also made in Europe, especially in Venice, in the later Middle Ages (e.g. Tait 1979; Baumgartner and Krueger 1988: 126–60; Tait 1999).

Metal plating and coloring

The metals widely used in the Middle Ages had a range of natural colors – gold, silver, copper and its alloys, lead, and iron – but objects made from them were often decorated with other metals to give a polychrome effect. The additional metal could be inlaid (mainly used on iron), soldered on, or applied as a plating, but only gold or silver plating involved any chemical transformations. Tinning (a thin tin or tin-lead coating) was widely used to stop the underlying iron or copper alloy corroding and was melted onto the cleaned metal.

For gold or silver plating the finely divided precious metal was ground together with mercury, which is a liquid at room temperature, to dissolve the gold and form a paste known as an amalgam (Bayley and Russel 2008). The amalgam was rubbed onto the cleaned surface of the object to be plated, which was then heated gently to drive off the mercury, and the plating was then burnished.

Copper was usually used as an alloy, and if the major additions were tin and/or lead the metal had a brownish hue rather than the pink/orange of the pure metal. However, if zinc was the main addition, then the alloy looked yellow and could almost be mistaken for gold. In the High Middle Ages the copper alloys often contained zinc, tin, and lead, but provided the main alloying element was zinc, the metal still had a yellowish hue. Making these golden alloys was not simple, as zinc metal was virtually unknown in the West at this time. Metallic zinc could only be produced by distillation – as was done in India and China (Craddock 1998) – because when zinc ores were smelted zinc vapor was produced at temperatures below those at which the zinc compounds were reduced to metal. In Europe and the Middle East brass was instead made – as it had been since Classical times – by heating together finely divided copper metal, zinc ores (usually the oxide or carbonate), and charcoal in a crucible, which sometimes had a lid to help retain the zinc vapor (Bayley 1998).[4] In the Islamic world the local zinc ores were sulfides so they had first to be roasted (calcined).

The zinc oxide was then heated and the fumes condensed, incidentally removing most of the impurities, so the brass produced had lower levels of impurities than that made from naturally occurring oxidized zinc ores (Craddock et al. 1998). Brass-making was a cementation process in which the charcoal gave a reducing atmosphere and zinc vapor sublimed from the ore and diffused into the solid copper; at the end of the process the temperature was raised and the copper (now brass) melted. Medieval texts consider brass-making as tincturing, changing the color of the copper. By heating the copper with an "earth" it changed color and also gained in weight.

TRADE AND TECHNOLOGY TRANSFER

The medieval period started with the fall of the Roman Empire in the West, which resulted in the loss of large-scale manufacturing activity. Despite this the Migration Period, with its movement of peoples and cultural change, shows some continuity, as in the supply of raw glass from the eastern Mediterranean to Northwestern Europe. Craft skills of the highest order continued in these peripheral areas. An example is gold jewelry inlaid with garnets that is found widely in Northwestern Europe – these gemstones originated in India (Mathis et al. 2008) and so demonstrate the existence of long-distance trade routes, though it is likely the goods changed hands many times as they traveled.

Two distinct types of soda-rich plant ash glasses – one from the eastern Mediterranean and the other from Mesopotamia – have been identified among the luxury Islamic relief-cut glass from the castle of Gauzón on the north coast of Spain. These reveal close commercial links in the tenth and eleventh centuries between the Christian kingdoms of Asturias and León on the Atlantic coast and the Islamic world of al-Andalus, and beyond it to Mesopotamia and the Levant (De Juan Ares et al. 2018).

The trade in soda ash from the Levant to Venice is well documented from the thirteenth century, and probably earlier, and its availability to the glassworkers there is one of the reasons for their dominance in the production of high-quality glass vessels. Levantine soda ash was also imported to elsewhere in Italy, but for use as an ingredient in soapmaking (Jacoby 1993). The varied uses to which high-lead glass was put once it appeared in both the Islamic world and in Northern Europe at the end of the first millennium CE argues against trade in the products, but its near-simultaneous appearance does raise the question – was it independently invented in the two areas or was its spread, perhaps from a single source, as a consequence of the Viking trade routes that linked the areas via the Volga?

Crucible steel was a valuable commodity made in multiple centers in the Islamic world and traded within it and to Europe, notably by Persian merchants. Williams (2007: 239) has argued that the renowned "Ulfberht" swords were produced from it. They have been found all along the trade route exploited by the Vikings in the ninth and tenth centuries that ran from the Middle East to the Baltic. Another commodity that traveled on this route was the silver coins of the Samanids (815–1005 CE), whose metal came from mines south of Samarkand and in Afghanistan (Merkel et al. 2013). After the fall of the Samanids in the eleventh century the trade along this route declined, as did the manufacture of "Ulfberht" swords, presumably because crucible steel was no longer available in the north.

Precious metals were also traded over long distances as they came from a limited number of sources. Newly smelted metal contains varying amounts of minor or trace elements, and their presence can be used to help identify its source. Before the mid-eighth century the gold from old coins was widely recycled – Byzantine and Sasanian coins in the East, Byzantine coins in North Africa, and Visigothic coins in Spain – but after this date the gold from Sudan (present-day Mali, Mauretania, and Niger) traveled north across the Sahara and gold dinars were struck in the mints located on the caravan routes, which then circulated in the Maghreb and in Spain and Sicily (Gondonneau and Guerra 2002). Evidence for gold refining, similar to that described above, and the production of coin blanks has recently been found in Tadmekka, to the south of the Sahara (Rehren et al. 2011; Rehren and Nixon 2014). From the twelfth century the Almoravids controlled this trade and their dinars were used in North Africa and al-Andalus but were also accepted as currency in southern France and across the Mediterranean (Messier 1974). In the Near and Middle Eastern mints, gold from Nubia, the Red Sea, or Yemen was used from the eighth century, but by the tenth century gold from another source was also being used – and by this time the eastern gold also seems to have reached the western mints (Gondonneau and Guerra 2002).

It is not only metals and glass that were traded over long distances. Mentioned above were some of the other high-value goods that traveled – silks, worsteds, and ivory. Other more everyday products such as ceramics are also found far from where they were made, though whether they were actively traded or just traveled with other goods, perhaps as containers for some of them, is more difficult to assess.

MEDIEVAL INDUSTRIES AS PRECURSORS TO LATER DEVELOPMENTS

Although many manufacturing processes display a high degree of continuity from earlier times, there were some changes that started in the medieval period and can be seen developing and accelerating in subsequent centuries.

Large-scale manufacturing had been seen in the Roman period, but in many areas there was a reversion to more local and smaller-scale production. However, this began to change, encouraged by the beginning of mechanization. Waterpower was used not only for grinding grain and fulling cloth but also to drive bellows and trip hammers. This development of machines continued to increase, growing in scale until steam-powered engines were developed in the eighteenth century. A by-product of mechanization was the ability to work on a larger scale: the early blast furnaces were larger than those that had gone before, but were themselves dwarfed by the increasing height and volume of postmedieval and modern furnaces. This innovation in ironworking produced liquid iron, and in quantities that were previously unthinkable, though it had to be decarburized before it could be used for manufacturing most types of products. The development of (re)fining processes was still to come, but led in the nineteenth century to Bessemer's Converter and the production of steel in vast quantities (e.g. King 2016).

The scale of production can be difficult to estimate, but heaps containing 4,000 m^3 of slag weighing well over 10,000 tons, found near Samarkand, were the by-product of the smelting operations that ultimately produced tenth-century silver dirhams (Merkel et al. 2013). The other development that the increased scale of production brought was the subdivision of labor. A single craftsman did not necessarily undertake every step in making an object. This can be clearly seen in mints where the chemical processes of ensuring the precious metal was of the right fineness were separate from producing the coin blanks and then striking them. In the tenth century there is evidence from York that all these processes were carried out on a single tenement (Bayley 1992), while by the sixteenth century the Mint in the Tower of London had separate workshops for each stage of the process (Bayley and White 2013).

The many changes in the composition of glass have been outlined above, and in Northern Europe this culminated in the change to potash glass, variants of which were used until the development of the Solvay process in the nineteenth century made artificial sodium carbonate cheaply available (Dungworth 2019).

SUMMARY AND CONCLUSIONS

Throughout the medieval period manufacturing flourished. Though the craftsmen's designs for individual products varied depending on the wishes and cultural affiliations of their customers, the chemical processes they used to make these products were generally similar across Europe, around the Mediterranean, and even further afield.

Medieval craftsmen were generally not innovators or experimenters. Once they had a process or chain of processes that produced the desired result they remained faithful to it. This conservatism is very obvious in the continuity of practices from Classical times, which often continued well into the early modern period and beyond. Despite this conservatism there were, as we have seen, some innovations that led to the development or adoption of new processes or products that quickly spread. One area where change can be seen is in the scale of operations. Increased demand led medieval industries to become more industrialized and mechanized and to operate on a larger scale.

In an age when travel was far more difficult than it is today, it can come as a surprise to see how similar the chemical industries in different areas were. The commercial innovations of the medieval period encouraged trade: rare and exotic materials and objects traveled long distances, some ending up in most unexpected places (e.g. the statue of Buddha that found its way to Helgö in Sweden in the eighth century and the garnets from India that were widely used in making early medieval gold jewelry in Northwestern Europe). The origins of the goods produced can be identified by modern scientific analyses, archaeological finds, and also from surviving medieval documents.

Some of these documents also contain descriptions of how things were manufactured, which complement the archaeological evidence for many of the chemical processes; by comparing the two we are able to fill out the picture of how things were made. This more detailed understanding helps us interpret the evidence and allows us to appreciate the complexity of the chemical processes routinely and successfully carried out in medieval times.

CHAPTER SEVEN

Learning and Institutions: *Teaching the Art East and West*

REGULA FORSTER AND JEAN-MARC MANDOSIO
WITH ANTOINE CALVET

ISLAMICATE WORLD[1]

Regula Forster

Alchemy as a secret art was not taught openly, or at least so the literary topos has it. Therefore, it is very difficult to establish how alchemical knowledge was acquired, taught, and transmitted. Physical evidence is scarce, and even that which exists cannot always be attributed with certainty to alchemical usage: for example, glass vessels found may have served an alchemical purpose, but they could also have been the property of a perfume-maker, serving as nothing more than vessels for mixing perfumes (Savage-Smith 1997). While for other arts and crafts we have inscriptions on products or instruments, similar testimonies are missing for alchemy – no signed vessels are known, and the products could not be signed for obvious reasons.

Literary forms

Arabic alchemy is of special interest to the historian of Arabic literature, as it uses genres not generally widely employed in Arabic, most importantly allegories, visions, and mythical narratives (Ullmann 1972: 145). Prominent are allegories of the seven metals, such as in *Kitāb al-Aṣnām as-sab'a* ("Book of the seven idols"), attributed to Apollonius of Tyana (Ullmann 1972: 173; Karimi Zanjani Asl 2013), and in *Risālat al-tāj* ("The epistle of the crown"), attributed to Mary the Copt, a concubine of the prophet Muḥammad (Holmyard 1927; Forster 2017: esp. 460). In *Kitāb al-aṣnām al-sab'a*, the seven idols of the title, each made from a different metal, give instructions, and are also portrayed as priests of the seven planets. The text thereby stresses the relationship between the planets and the metals. However, gold and silver are given a special status, as they are each treated in two sections, so that the whole work consists of nine rather than seven chapters. Likewise, *Risālat al-tāj* sets gold and silver apart. Here Mary tells a story about a mother who has seven sons. Five of them complain about the fact that two of the siblings have become powerful kings, while they themselves have remained lowly. Their mother explains that they had been born in a time when she was either too young or too old. The only method of gaining the same status as their siblings, she asserts, is to return to her womb and to be born again. This makes clear that the seven brothers are to be interpreted as the seven metals, the two kings among them being gold and silver, while the mother is earth. In what follows, the text presents an allegorical dialogue, ending in a description of the Ouroboros.

An allegorical vision is attributed to the Greek philosopher Archelaus, who is also considered to be the author of *Musḥaf al-jamā'a* ("The book of the group," better known under the Latin title as *Turba philosophorum*): in *Risālat madd al-baḥr dhāt al-ru'yā* ("The visionary epistle about the sea's rising," translated into Latin as *Visio Arislei*), the fusion of sulfur and mercury into cinnabar is presented as an act of procreation (Ullmann 1972: 153). In the much later *Risālat al-ruhāwiyyāt* ("The letter of the Edessian revelations") by one al-Marrākushī (fifteenth century?), the author presents a dream about a beautiful Persian youth, whom he desires, follows to Edessa (now Urfa), and marries off to an Egyptian slave girl, etc., therewith symbolizing the alchemical process (Ullmann 1972: 245).

Besides these genres not generally found in Arabic literature, more common genres abound, especially treatises, commentaries, dialogues, and didactic poetry. The latter may take the form of shorter or longer poems in the so-called *'arūḍ* meters or are written in the traditional meter of didactic poetry, the short *rajaz* verse. A large collection of alchemical poetry, entitled *Firdaws al-ḥikma* ("The paradise of wisdom"), is attributed to one of the alleged founding fathers of Arabic alchemy, Khālid b. Yazīd (d. ca. 704), an Umayyad prince (Ullmann 1972: 193–4).

Later, the Shiite Ibn Umayl al-Tamīmī (fl. tenth century) composed a long *rajaz* poem, *Risālat al-shams ilā l-hilāl* ("The epistle of the sun to the crescent moon"), on alchemy. As is not uncommon with didactic poetry, the verses were considered so difficult that the author himself composed a commentary on his verses, the famous *Kitāb al-mā' al-waraqī wa-l-arḍ al-najmiyya* ("The book of the silvery water and starry earth," ed. Stapleton et al. 1933). The commentary was later translated into Latin (Ullmann 1972: 217–20).

From the twelfth century, two important collections of alchemical poetry are known, one by al-Ṭughrāʾī (d. 1121), who was a civil servant of the Saljuqs besides being a theoretician of alchemy (Ullmann 1972: 231), and the other by the Moroccan Ibn Arfaʿ Raʾs (fl. twelfth century), entitled *Shudhūr al-dhahab* ("The splinters of gold"). Consisting of around 1,400 verses and more than forty poems, the collection comprises allegorical poems in *rajaz* and the *ʿarūḍ* meters. Ibn Arfaʿ Raʾs was highly praised not only as an alchemist, but also as a poet, as biographers call him "the poet of the sages and the sage of the poets" (al-Kutubī in ʿAbd al-Ḥamīd 1951–3: II: 181; al-Ṣafadī in Baalbaki 1983: XXII: 260). *Shudhūr al-dhahab* survives in about 100 manuscripts and was the object of commentaries for centuries. The tradition of commenting on *Shudhūr al-dhahab* was started by Ibn Arfaʿ Raʾs himself, who wrote a *Ḥall mushkilāt al-shudhūr* ("Solution to the problems of 'The splinters'"),[2] and was continued by many later alchemists, including al-Sīmāwī (fl. mid-thirteenth century), al-Jildakī (fl. mid-fourteenth century), and ʿAlī Bek al-Iznīqī (fl. fifteenth or sixteenth century), also called *al-muʾallif al-jadīd* ("the new author").

Several alchemical works were attributed to women, or women were portrayed as adepts of the divine art. Women appear to have been seen as likely authors of alchemical writings and as feasible practitioners. However, this was true only when looking back: no famous women of Islamic times were ever mentioned as alchemists; the latest woman to have been credited with the authorship of a treatise on alchemy was Mary the Copt (who was a concubine of the prophet Muḥammad, but is famously confused with other Marys, namely Moses' sister and Jesus' mother; see Ullmann 1972: 181–3).

Finally, in keeping with the topos of alchemy as a secret art, not to be taught to the unworthy, there is yet another topos – namely that alchemical texts are hard to find. Authors tell the reader of their arduous search for a book, a tablet, or the like, often in the ground, in an ancient temple, or in a ruin, and of the extreme difficulty of reading the texts in question, often written in foreign scripts and languages (Ullmann 1972: 166–7, 219, 222; Forster 2006: 52). A typical example may be found in the introduction of *Risālat al-sirr* ("Epistle of the secret"), an early Hermetic treatise:

> This epistle was found in the inner Akhmīm, under a tablet of marble, in a tomb in which there was a dead woman of perfect physical constitution. Her plaits extended to her feet. On her lay seven gilded dresses, and each had a golden button. Around her were small beds on which were dead people who looked like young men. And this epistle was under her head, (written) on a tablet of gold, looking like an enormous shoulder blade. It was written

> in black in a script that we have shown at the end of the book. This was in the time when al-Ma'mūn [Abbasid caliph, r. 813–833] was in Egypt. It was interpreted for him together with the psalms, which were interpreted as we have explained already. And the interpreter was a man from Ḥimyar, who was learned in the scripts.
>
> (ed. Vereno 1992: 137, §§ 2–9; translation RF)

The usage of this topos emphasizes the importance of alchemical knowledge and of the content of the book presented to the reader.

Oral transmission

As oral transmission of knowledge is seen as the ideal form of teaching and learning in the medieval Islamicate world (Rosenthal 1947; Hirschler 2012: esp. 11–31; Schoeler 2013), it seems likely that alchemy was taught orally, by a teacher or master, to their student or students. However, evidence for such teaching situations is scarce. There is no formal system of teaching certificates (*ijāza*) in alchemy, as in other fields of study, especially in the teaching of the sayings of the prophet Muḥammad, where a student would only be allowed to teach the sayings he had heard from his teacher, once he would have acquired the relevant *ijāza* from his teacher (Günther 2017: esp. 38).

That alchemical texts were, like those in other fields of knowledge, sometimes not simply copied from manuscripts, but copied out following dictation (Günther 2017: 35), hence a form of oral interaction presumably by a teacher, can be deduced by the slips in the textual transmission that sometimes occur. In the transmission of the poetry collection *Shudhūr al-dhahab* by Ibn Arfaʿ Ra's, there are several instances of such slips, for example *tā'* instead of *thā'* or a confusion of *ẓā'* and *ḍād*. Dictation was also probably the reason for deviations such as *akāmū* ("they piled up") instead of *aqāmū* ("they set up") and missing articles (*bi-nūr*, "through light," instead of *bi-l-nūr*, "through the light," or *li-sirr*, "for secret," instead of *li-l-sirr*, "for the secret").[3]

Furthermore, many texts on alchemy were written in the form of literary dialogues, thereby stressing the importance of oral instruction: the dialogue as a literary genre is the written representation of oral interaction that is supposed to have taken place (for a definition of the genre, see Forster 2017: 2–7). It remains difficult to assess whether these texts represent reality or rather an ideal not usually achieved, but in either case, they show that alchemy was perceived as an art that should be learned in a face-to-face setting.

Dialogues were already used in Greek alchemical writings (see Berthelot 1887: II:56–69, III:289–99; Hallum 2008: 242). Islamicate alchemy appears to have inherited the form and took it up over and over again. Among the oldest Arabic dialogues on alchemy are texts attributed to Zosimos of Panopolis (fl. fourth century), perhaps translations from Greek, but more likely Arabic compositions based on genuine Greek material attributed to Zosimos (Hallum 2008: esp. 257; contra Abt 2007b: esp. 21–68).

Furthermore, the alleged founding father of Arabic alchemy is credited with a dialogue: *Masā'il Khālid li-Maryānus al-rāhib* is a dialogue between the Umayyad prince Khālid b. Yazīd and the Christian monk Maryānus. Though it is unlikely that the historical Khālid was interested in alchemy, and *Masā'il Khālid li-Maryānus al-rāhib* therefore should be considered to be pseudepigraphic, the corpus of alchemical texts attributed to Khālid is quite extensive and his fame as an adept of the art spread widely (Ullmann 1978; Anawati 1996: 864; Dapsens 2016).

The dialogue form was adapted by many (mostly anonymous) authors, who presented prominent dialogue partners, like Cleopatra (d. 30 BCE), Thābit b. Qurra (d. 901), or Ibn Waḥshiyya (fl. ninth to tenth centuries CE; Forster 2017: esp. 41–4). It is quite striking that female dialogue partners – already present in the works attributed to Zosimos, who talks to his pupil Theosebeia – appear in many dialogues. This might suggest that alchemy was seen as at least a not inappropriate field of learning for women. However, it could also be due to the fact that in presenting a male and a female dialogue partner, the male and the female principle central to alchemy would be present (Lippmann 1919: esp. I:80, 99). The usual arrangement is a dialogue between master and disciple. Few texts differ from this scheme, either by presenting a group of students (who would, however, usually speak in one voice only) or by introducing two teachers, like *Risālat al-ḥakīm Qaydarūs* (Forster 2017: esp. 58–9). The dialogue between Aristotle and the Indian sage Yūhīn found in *Kitāb fīhi khabar Yūhīn al-hindī* ("The book containing the story of the Indian Yūhīn," ed. Müller 2012: 13–26) is an exception, as it does not present a teacher–pupil relation. Rather, it shows a friendly discussion between two scholars from different cultural traditions (Forster 2017: esp. 80). Finally, an interesting case is *Muṣḥaf al-jamā'a*: this is not so much a dialogue, but rather the text presents a whole group of sages evaluating the secrets of alchemy in a *majlis*-like setup. The *Muṣḥaf al-jamā'a* is also an interesting text as its Arabic original is only preserved in fragmentary form, while a complete Latin translation, the *Turba philosophorum*, is extant (Ruska 1931; Rudolph 1990).

The importance of oral instruction also becomes clear when considering the coded language usually employed. These codes, in scholarship most often called by the German term "*Decknamen*," are used in most alchemical texts, not only in those of a more allegorical strand. Some of the codes have become more or less general, and the designated substances are therefore easy to detect. For example, *shams*, "sun", is used for gold, *qamar*, "moon", for silver, and *'uqāb*, "eagle," for sal ammoniac. Other usages of *Decknamen* are very limited, be it to a certain author or even a certain work. Lists of such code names may be found in several works, such as in al-Jildakī's *Kitāb al-burhān* ("Book of the proof"). However, these lists never became a literary genre of their own. The modern *Decknamen* compilation by Siggel (1951) should be used cautiously, as its sources are not always clear.

Oral instruction in a laboratory context might be the context for yet another genre of alchemical writing: the recipe. Recipes are usually preserved in very much abbreviated forms so that they can be understood only by the initiated (Braun 2016: 35). It is also important to note that some elements of the actual physical preparation one could call tacit bodily knowledge are never transferred to the recipe text: one will not learn how exactly to hammer, to stir, etc. (Moureau and Thomas 2016). The physical appearance of recipe texts is also revealing. Usually, recipes were written on flyleaves and added to larger manuscripts on spare leaves (Margoliouth and Holmyard 1931; Braun 2016: 53). This seems to indicate a usage in a laboratory or practical context rather than one of book learning. However, larger collections of recipes are known to have existed. One of them, probably dating from the tenth to eleventh centuries, has been – most probably wrongly – attributed to the rationalist judge 'Abd al-Jabbār al-Hamadhānī (d. 1024). That even a judge could be depicted as an alchemist is noteworthy (Leube 2013). Furthermore, recipes are presented as an easy way to achieve alchemical success by the fourteenth-century magician and alchemist Muḥammad b. al-Ḥājj al-Tilimsānī (Ullmann 1972: 234).

Textual transmission and manuscripts

Though the importance of oral instruction is repeatedly emphasized, learning from books seems to have been well accepted (differently from other fields of scholarship). The importance of books is already stressed in the Jābirian corpus, when students of alchemy are reminded that studying books is an essential part in acquiring alchemical knowledge:

> We say: God grants success! What the student of this art (*ḥikma*, literally wisdom) needs more than anything else is patience (*ṣabr*) for studying the sciences, then our books about them. ... I say in addition: The reader of these our 112 books – and one said: these our books – must not read them in the wrong order or in an unstructured order of reading or without a master (*ustād*). For the truth will not become manifest to whoever reads our book or our books in a different way. Rather, he should, on his part, be gentle towards his master. If he cannot find a master on them (i.e. the books), he should study them with those of his companions who have more understanding than himself or more learning in the neighboring sciences, such as medicine, philosophy, and logical reasoning. If he does so, truth will become manifest to him.
>
> (Holmyard 1928: 100–1; translation RF)

In a text that may date to the ninth to tenth centuries, *Risālat al-ḥadhar* ("The epistle of caution"), attributed to the Greek sage Agathodaimon, the importance of alchemical books is emphasized even more strongly. The dying Agathodaimon instructs his pupils:

> This discourse of mine may now be considered to be of sufficient length, because if you do not carry out all the processes of the Work, but (try to) avoid (the details of) its knowledge, death will be better for you than the sight of any book. Search therefore the books of the Sages, for in them the Work is hidden.
>
> (MS Oxford, Museum of the History of Science, Ms Stapleton 15: fol. 11r; translation by Stapleton from MS Oxford, Museum of the History of Science, Ms Stapleton 109 [typescript]: 26–7)

Indeed, books, mainly in manuscript form, must have played an important role in the transmission of alchemical knowledge in the Islamicate world: there is an extraordinary number of alchemical manuscripts in Arabic, Persian, and Ottoman-Turkish preserved in libraries all over the world. Some of these collections are accessible through specialized catalogs that reveal the importance of alchemical books (Siggel 1949–1956; Ullmann 1974–1976). However, as with most sciences, manuscripts, even of early texts, tend to be late, stemming from the seventeenth to eighteenth centuries, with some even later examples from the nineteenth and twentieth centuries. Despite the richness of the material available, ownership and patronage of these manuscripts have not yet been studied on a larger scale. For Ottoman Damascus, Liebrenz has shown that books on the occult sciences and especially on alchemy were more prominent in private collections than in endowed public libraries, and that members of the military elite did not have a very expressed interest in these sciences (2016: esp. 211, 319). It seems likely that the situation was not very different in earlier times.

Places of teaching and learning

Generally speaking, informal learning settings and established learning institutions are known to have existed in the medieval Islamicate world, and there is one entity that could be compared to the university, namely the *madrasa*, an institution where Islamic law and related fields were taught from the early tenth century onwards. The *madrasa* as a mainly religious institution seems to be an unlikely place for the transmission of alchemical knowledge. Alchemy, as an art and science, would obviously not be at the core of the curriculum. However, manuscript evidence proves that the *madrasa* milieu was not intrinsically opposed to alchemy and that manuscripts were indeed produced in such contexts or kept at such institutions. Therefore, we can assume that even in such a milieu, alchemical texts were read, probably for their philosophical contents, even though it is unlikely that any *madrasa* established a laboratory for practical work. A relatively early example is a manuscript today kept in London (MS London, British Library, Or. 13006), a collection of alchemical writings, parts of which were copied at the Madrasa al-Mawlawiyya al-ʿAlawiyya in Fez in 919 AH/1513 CE (see fol. 11r).[4]

In contrast to alchemy, the instruction in a somewhat comparable art, namely medicine, has been much more thoroughly studied. The parallel is a useful one, as we can assume that to become an alchemist, like a physician, one had two principal options: to learn with a master or to teach oneself (Pormann and Savage-Smith 2007). Since learning on one's own would be problematic, given the coded language, learning with a master would be the better option. References to a master are therefore quite common in alchemical writings, starting already in the Jābirian corpus, where Jābir more than once refers to his teacher in alchemy, Jaʿfar b. Muḥammad, called his *sayyid*, his master (Haq 1994: esp. 15). Traditionally, Jaʿfar is identified as being Jaʿfar al-Ṣādiq, the sixth *imām* of the Shia, but other identifications had been offered already in medieval times (Sayyid 2009: II/1:450–8). In the twelfth century, ʿAbd al-Laṭīf al-Baghdādī mentions his teacher in alchemy, though he, looking back, condemned alchemy as being useless (Allemann 1988: 13–15). In the fourteenth century, al-Jildakī traveled to Syria, Iraq, Byzantium, the Arabian Peninsula, and the Maghreb in order to learn from masters, before eventually meeting his principal teacher (Brockelmann 1937–1949: II:138–9, S II:171–2; Holmyard 1937: 47).

While medicine was taught at the hospitals from the tenth century onwards, there was no institutional place to learn alchemy; it was therefore presumably taught informally, in study circles and assemblies typical for all kinds of education in the medieval Islamicate world, taking place in mosques, courts, and private houses (Günther 2017: 31). However, evidence for the actual informal teaching of alchemy remains scarce. Leo Africanus (ca. 1490–1550) recounts that the Great Mosque in Fez was home to a study circle of alchemists, who were reading older works (Amadori 2014: 315). Therefore, the mosque must be considered if not an alchemical institution, at least a place where alchemists could meet informally. Furthermore, we can assume that artisanal chemistry no doubt was taught in the context of dyeing, glassmaking, pharmaceutical alcohol distillation for medical purposes, etc.

An interest in such practical alchemy must have existed at several courts. Princes, caliphs, and sultans alike seem not only to have sponsored the production of literary works and manuscripts, but also to have set up laboratories.

While any involvement in alchemy of the Umayyad prince Khālid b. Yazīd seems at least doubtful, we might assume that the Abbasid caliph al-Manṣūr

(r. 754–775) would have set up an alchemical workshop. He had sent an ambassador to the Byzantine emperor Constantine V Copronymos (r. 741–775), who saw there a demonstration of the functioning of the elixir. The caliph, so the legend has it, was extremely impressed and ordered that this art should now be searched for (Strohmaier 1991).

Practitioners were quite often found at courts, as rulers hoped to fill their empty treasuries: for example, the Mamluk sultan Qānṣawh al-Ghawrī (r. 1501–1516) is famous for having spent a large amount on a (obviously not successful) alchemist (Johnson 1996: 1).

Though an interest in alchemy was attributed to several rulers, princes, and caliphs (see Chapter 5), manuscript evidence for patronage remains scarce for the medieval period. The Almohad sultan Yaʻqūb al-Manṣūr (r. 1184–1199) of Morocco is called an alchemist in two manuscripts and also in legends connected to his name (Aït Salah Semlali 2015: esp. 166–9). Furthermore, in the east of the Islamicate World, the Timurid prince Jalāl al-Dīn Iskandar b. ʻUmār Shaykh (d. 1415), who was an active patron of culture, seems to have had a deep interest in alchemy: he ordered the composition of an alchemical work in Persian, entitled *Āʼīna-yi Sikanadarī* ("Alexander's rule"), and he was the dedicatee of a *Risāla-yi kibrīt-i aḥmar* ("The epistle of the red sulfur," Storey 1977: 437, no. 760–1).

LATIN WORLD

Jean-Marc Mandosio

Given the scarcity of archaeological evidence (see Chapter 3), our access to medieval alchemy rests almost entirely upon the written tradition. However, one should proceed with extreme caution when dealing with this notoriously cryptic and deceptive literature. If its theoretical aspects are well documented (see Chapters 1 and 4), the channels through which alchemical practices were actually learned and carried on are not so clear, and the matter is still open to discussion. The predominance of texts and, to a lesser extent, of images in our documentation may generate a bookish bias that could lead us to minimize other ways of learning and transmission, about which we can only formulate plausible conjectures.

Textual transmission

Except for a few isolated recipes such as those found in the ninth-century *Mappae clavicula* (Halleux and Meyvaert 1987) and other early medieval "books of secrets," the textual transmission of alchemy in the West really began with the twelfth-century translations of Arabic works into Latin. The literary forms inherited from the Greek–Arabic tradition – dialogues, allegoric visions, revelations, recipes – remained typical features of the medieval alchemical literature written in Latin and, from the late thirteenth century onwards, also in vernacular European languages. A landmark in the reception of Arabic alchemy was the translation and adaptation of *The Assembly of Philosophers* (*Turba philosophorum*, ed. Lacaze 2018), a compilation of sayings ascribed to a host of Greek philosophers gathered under the guidance of Pythagoras by his disciple Arisleus. This work, together with the *Emerald Tablet*, the *Book of Seven Chapters*, and other influential writings ascribed to Hermes Trismegistus, helped establish the antiquity and nobility of alchemy, which was not perceived by Latin scholars as an Arabic novelty but as a discipline both deeply rooted in Greek science and divinely inspired.

For several works extant in Latin to which no Arabic counterpart has been found, it is difficult to ascertain whether they are translations or not (Moureau 2020), a good case in point being the pseudo-Jābirian Latin production. Perhaps even more importantly, what pass for original Latin works are often, for the most part, patchworks of quotations – as were many Greek and Arabic treatises before them – so that modern conceptions of originality and plagiarism hardly apply to this literature.

European alchemical writers added a few literary innovations to the Arabic heritage. Being Christian, they elaborated their own religious allegories, notably the representation of the philosophers' stone under the guise of Jesus, whose miracles – healing the sick, multiplying food, resurrecting the dead, and ultimately rising from the grave and ascending to heaven, as well as the recurring transformation, through the Eucharistic rite, of bread and wine into Christ's flesh and blood – served as images of the stone's properties, while the assumed reality of transubstantiation became the standard theoretical ground for the possibility of metallic transmutations. Another innovation was the production of comprehensive alchemical works mimicking the form of scholastic treatises, the most remarkable ones being the late thirteenth-century pseudo-Jābirian *Compendium of the Achievement of the Great Work* (*Summa perfectionis magisterii*, ed. Newman 1991) and Petrus Bonus of Ferrara's *New Precious Pearl* (*Pretiosa margarita novella*, ed. Manget 1702: 2:1–80), dated 1330.[5]

An important difference between Arabic and Latin cultural contexts is the rise of universities from the end of the twelfth century onwards. The natural philosophy curriculum at the Faculty of Arts was centered on the Aristotelian corpus; therefore, as a result of the addition of Avicenna's chapters on minerals to Aristotle's *Meteorologica* (see Chapter 4), the "question of alchemy" (*quaestio de alchimia*) became a standard topic. However, it should be pointed out that the "question" usually consisted of a discussion on the generation of minerals, the physical foundations of alchemy, and the possibility of transmutations. The aim, of course, was not to teach the art of alchemy, which had no more reason to be taught in medieval universities than any other "mechanical art" (on this concept, see Chapter 4).

Interpretive issues

The Great Work was considered extremely difficult, if not impossible, to achieve, but one task that was nearly as difficult was to make a coherent sense out of the textual sources of alchemy. Their most striking feature, a characteristic that distinguished alchemy from the other medieval learned disciplines, was the deliberate obscurity of its language and the ambiguity of its concepts, which made the comprehension of alchemical writings a challenge for the alchemists themselves. There are innumerable instances of anxious confrontation with the riddles transmitted by earlier works, to the puzzlement and despair of their readers.

Virtually every aspect of the alchemical lore was controversial, beginning with the meaning of words. There was hardly a shared view of the alchemical vocabulary, even though *synonymiae* (i.e. glossaries) had been devised to help readers and practitioners find their way through the maze of coded phrases and borrowings from Arabic. The most famous of these glossaries, a work probably compiled in the fourteenth century and falsely ascribed to the thirteenth-century grammarian John of Garland, was published in 1560 under the title *Explanation of Synonyms in the Alchemical Art* (*Synonymorum in arte alchimistica expositio*, in Mandosio 2001). The compiler gave mostly correct translations of Greek and Arabic words and took into account multiple Latin transcriptions or distortions of those words: for instance, *adebessi*, *dabestis*, and *rebis* are three variants of a same unidentified word meaning "turtle," *alexir* and *elexir* (*al-iksīr*) are both registered as meaning "medicine," and so on. Conversely, we may also find in Pseudo-John of Garland's glossary up to seven definitions for a single word, as is the case for *alkali* (var. *alcali*, *algali*). One should be aware, though, that they might actually refer not only to the same word but to the same substance; for instance, the words *calcantum*, *calcanthum*, *calcatus*, *calcadrium*, and *calcure*, all variants of the Greek word *chalcanthos* (lit. "copper flower"), are defined in the glossary either as "burned copper" (*aes ustum*) or "vitriol," but these are two appellations of copper sulfate, also known as blue vitriol.

However, it was not uncommon for a variant spelling to be interpreted as an entirely different word, whose meaning had to be guessed or reconstructed by readers according to the context and to their own knowledge of the matter. This process could also be originated by the loss of the original meaning of Greek or Arabic words, as in the case of *telesmus*, a Latin neologism that appeared in the "vulgate" version of the *Emerald Tablet* ascribed to Hermes Trismegistus (Mandosio 2004, Mandosio 2005). It was a transcription of the Arabic word *tilasm*, meaning "talisman" or "charm," itself derived from the Greek *telesmos*. It was correctly translated as *praestigium* in Hugo Sanctellensis' twelfth-century version of the *Tablet*, included in the Arabic *Book of the Secret of Creation* (*Kitāb sirr al-khalīqa*, ed. Weisser 1979; *De secretis naturae*, ed.

Hudry 1997–1999). But the anonymous translator of the "vulgate" version (see Chapter 4) apparently did not understand the meaning of the Arabic word and merely Latinized it as *telesmus*. From then on, a notable semantic drift began. An early Latin commentary on the *Emerald Tablet* entitled *Explanation of the Words of Hermes, Master of Philosophers, According to Our Truth* (i.e. the Christian truth; *Expositio verborum Hermetis magistri philosophorum secundum veritatem nostram*, ed. Steele and Singer 1927) explained that "among the Arabs, *telesmus* means divination," and the anonymous commentator drew from this wrong interpretation his conviction that the mention of *telesmus* by Hermes was an allusion to some "secret" – probably because a secret is that which has to be divined. In his wake, the fourteenth-century commentator of the *Tablet*, Hortulanus (ed. Ruska 1926), declared that *telesmus* meant "secret" or "treasure," and that in the alchemical context it specifically referred to "the Stone" and, more precisely yet, to "the final way in the work of the Stone."

Another unsettling feature was the polysemic nature of ordinary words or names mentioned in alchemical contexts. For instance, Pseudo-John of Garland's glossary stated that the name "Mars" (i.e. iron) could also be used to designate "Venus" (i.e. copper). Medieval readers were regularly warned that, in alchemy, any word could refer to something quite different from its standard meaning. This depressing state of affairs was confirmed, rather solemnly, at the very end of Pseudo-Jābir's *Summa*, when the author declared that he had "hidden [the science] where [he had] spoken more openly," even though he did not hide it "under enigma" but "with a plain sequence of speech." The question, then, was to determine whether, in a given passage, words and phrases were to be taken literally or with a grain of salt.

A telling example of the kinds of dilemma generated by the semantic instability of the alchemical language was the interpretation of the role played, in treatises and recipes, by organic substances, extracted from the bodies of animals and plants (Mandosio 2012): should they be used as such, or were they merely code names for mineral substances? And what to make of that famous oxymoronic statement from the long version of the pseudo-Aristotelian *Secret of Secrets* (*Secretum secretorum*), translated from Arabic by Philip of Tripoli in the years 1230–1240 (Pseudo-Aristotle, ed. 1555): "Take the animal, vegetable and mineral stone, which is not a stone and does not possess the nature of a stone," whose "proper name" is the Egg, "that is to say the Philosopher's Egg"? Every reader had to elaborate their own understanding of the "stone which is not a stone," or to pick up an already extant interpretation fitting their own conception of the Great Work. There was virtually no end to the confrontations raised by issues such as these.

The absence of a stable paradigm – that is, a general agreement on even the most basic concepts – makes it tempting to view the medieval alchemist as "a man alone in front of a book" (Halleux 1979), and the alchemical literature itself seems to confirm this assumption. No small part of it was dedicated to the compilation and interpretation of alchemical sayings borrowed from ancient and prestigious authorities, a feature that supports the idea that the alchemical lore was mostly of a bookish nature. But it was not bookish in the same manner as, say, medieval logic; for, as the *Summa* warned, "he who inquires by the pursuance of books will arrive very slowly at this most precious art." As a matter of fact, the vast majority of alchemical texts, including those explicitly presented as "explanations" (*expositiones*), were not really aimed at explaining things thoroughly (Mandosio 2000). The alchemists liked to compare their written works to labyrinths, and their commentaries were for the most part circular argumentations based upon authoritative statements. Every new work presented itself as a purveyor of the "true" alchemical dogma, which the author claimed to have untangled from carefully selected authorities.

There were, however, recognizable branches or families of works sharing common features. They could emanate from groups of authors belonging to the same religious or courtly milieu, though their doctrinal kinship could also well be the result of a purely bookish influence. The reconstruction of the chronology and authorship of alchemical literature is hampered by the mystifying view of the history of alchemy maintained by the alchemical writers themselves. This view is characterized by the common belief that the most renowned alchemical works were more ancient than they actually were, a belief that was continually reinvigorated by an abundant production of pseudepigraphical writings.

Pseudepigraphy

The European alchemists inherited from their Greek and Arabic predecessors the habit of ascribing new works to renowned authorities, ancient or modern. The birth of a medieval pseudepigraphic tradition is aptly illustrated by the fate of the French poet Jean de Meung (second half of the thirteenth century). He became famous for writing the second and lengthiest part of the *Romance of the Rose*, which became in his hands a *Mirror for Lovers* dealing with all sorts of subjects (Mandosio 2017). It contained eighty-four verses in praise of alchemy, described as the only human art capable of emulating nature (transl. Horgan 1994). The poet was the first French author to write about alchemy in the vernacular; the great fame of his work and the prominent status of these verses inspired several alchemical poems bearing Jean de Meung's name (Calvet 2017). In this case, the corpus of alchemical works ascribed to him was motivated by his explicit appreciation of the art. In the same manner, Roger Bacon (d. 1294), who also praised alchemy (see Chapter 4), unwittingly became one of the most renowned alchemical medieval writers, thanks to a series of works composed by admirers under his name.

The attribution of numerous alchemical works to Albert the Great (d. 1280) arose from a slightly different situation, for Albert had dealt with alchemy, in his *Meteora* and *De mineralibus*, in a rather cautious manner (Halleux 1982). The unknown writers who composed these works did so in order to make him appear as an outspoken promoter of alchemy (Calvet 2012). The same could be said for Thomas Aquinas (d. 1274), whose position on alchemy was equally cautious: metallic transmutations are theoretically possible but hardly feasible in practice. Accordingly, he also became after his death a defender of alchemy. The contradiction between his authentic works and the alchemical writings going under his name was explained quite simply in the fourteenth-century *De esse et essentiis* (ed. Rigius 1488), a pseudo-Thomistic work whose title echoed Thomas's *De esse et essentia*: the Pseudo-Thomas declared that for a long time he was hostile toward nigromancy and alchemy, but that in his old age he discovered that those sciences were true. Therefore, his last word on the matter should be viewed as more decisive than his earlier position. This "testamentary" topos is a recurring feature of alchemical literature, the most famous example being the fourteenth-century *Testamentum* ascribed to Ramon Llull (ed. Pereira and Spaggiari 1999).

Both Albert and Thomas were Dominicans, and it is likely that the alchemical corpora attached to their names were mainly elaborated by members of their own religious order who practiced alchemy and were embarrassed by the fact that such leading figures did not share their enthusiasm for this activity. The same goes for the Franciscans, who ascribed dozens of alchemical works to Ramon Llull after his death in 1315 (Pereira 1989). He saw alchemy as a fraud, and this was obviously unacceptable for his fellow brothers, who were as keen as the Dominicans to legitimate their alchemical practices against recurring condemnations (see Chapter 5). A good way to challenge the validity of such condemnations was to show that the most eminent members of the order were in fact very skilled alchemists. This is why the literary fiction of the "testament" was successfully applied to Llull, who was considered for centuries as an alchemist, in full opposition to his real views. The famous physician Arnald of Villanova (d. 1311), who seems to have had little interest in alchemy, also became posthumously the author of an ample alchemical corpus (Calvet 2011).

The production of pseudepigraphical alchemical works was so massive and successful in late medieval and early modern times that those who pass, even today, for the most famous alchemists of the Middle Ages, namely Albert the Great, Arnald of Villanova, and Ramon Llull, never wrote a single alchemical treatise. Moreover, two other renowned "medieval" sources – the fourteenth-century French alchemist Nicolas Flamel and the fifteenth-century German monk Basilius Valentinus – were not medieval at all: the first was a real person who never practiced alchemy and whose legend began in the sixteenth century (his famous fictitious autobiography, the *Book of Hieroglyphic Figures* [*Livre des figures hiéroglyphiques*] was published in 1612), and the second was invented by a Paracelsian called Johann Thölde in 1604, who wished to add historical prestige to the medicinal use of antimony by feigning that his *Triumphal Chariot of Antimony* (*Currus triumphalis antimonii*) had been written two centuries earlier by "Brother Basilius Valentinus."

In addition to the topos of the late conversion of a wise man to alchemy, the typical pseudo-biography of an alchemist comprised several mandatory features. First, the protagonist desperately tried to make sense out of one or more books written in coded language; then he passed a rather long time going from one mistaken practice to the other, listening to bad advice given by unreliable masters, until he met a true adept who gave him the "key" to the correct practice of alchemy, or he was granted the blessing of divine inspiration; and finally, after having achieved the Great Work, he decided to put his personal experience into words, as his predecessors had done, but cryptically enough to be understood only by those worthy of it while remaining incomprehensible to others. Even when the authors did not hide behind false identities, the tales of their experience inevitably entered the realm of fiction or delusion as soon as they pretended to have performed real transmutations or extraordinary medical cures by alchemical means.

Practical transmission

In a context that made history barely distinguishable from myth and fiction,[6] it is difficult to reconstruct with some plausibility how alchemy really was taught. There undoubtedly was a great deal of laboratory practice that could not be learned merely in a bookish way: even though the instruments, mineral substances, and basic operations such as distillation, sublimation, and calcination were accurately described in works such as the *Summa perfectionis magisterii*, an empirical apprenticeship was necessary to perform actual experiments. The practitioner of alchemy was likely to be a person acquainted with the techniques of pharmacy and/or metalworking, while the typical alchemical writer was either a physician or a cleric trained in natural philosophy at the Faculty of Arts. In the pseudo-Jābirian *Summa*, probably written by the Franciscan Paul of Taranto (Newman 1991), the perfect alchemist was described as a man who had to be "learned and advanced in the knowledge of natural philosophy." In his *New Precious Pearl*, Petrus Bonus of Ferrara labeled himself as a physician, and the author of a famous fourteenth-century commentary on the *Emerald Tablet*, Hortulanus, claimed to be a Franciscan. It may reasonably be surmised that not all alchemical writers were consummate practitioners. It would seem that the more allegoric a work was, the more remote it was from laboratory practices, and conversely, the works that consisted mainly of recipes reflected more closely what the practitioners of alchemy really performed or tried to perform in their laboratories. But collected recipes were often copied from earlier collections and therefore were no less bookish than works of a more theoretical or allegorical nature. It should be observed that, due to a common philosophical or literary bias and to the lack of technical capacity, many scholars tend to neglect the rich manuscript transmission of recipes, which are seldom edited and remain understudied – one rare exception being the excellent edition of the *Sedacina*, a fourteenth-century compilation of recipes (ed. Barthélémy 2002).

In medieval Europe, alchemy was not a trade that could be learned through institutionalized apprenticeship, since there was no such thing as a guild of alchemists. The accounts we may find in alchemical literature on the subject cannot be taken at face value. They typically pictured, on the one hand, a crowd of ignorant or illiterate, self-taught, and misguided practitioners who experimented haphazardly with all sorts of substances and were unable to achieve anything but delusion and fraud, while on the other hand a very small elite of adepts, capable of understanding how to put into practice the doctrines that the wise men of the past transmitted in their cryptic writings, had access to the Great Work. This conception was obviously biased by the belief that there actually was a definite way of interpreting alchemical writings in order to achieve the Great Work; consequently, the authors of such writings tended to ascribe the ineffectiveness of the alchemical processes that should lead to the transmutation of metals to the inadequacy of ordinary practitioners rather than to the foolishness of the goal they pursued.

THE TRANSMISSION OF ALCHEMICAL LEARNING

Antoine Calvet

Not much is known about the education of alchemists. It is assumed that the master alchemist, who was a keen student of nature, trained one or more workers in the laboratory and instructed them in the procedures and use of the apparatus, the furnaces, and the still. Our only witnesses of an apprenticeship, which was mostly carried out *viva voce*, are recipes, some texts of didactic scope such as the *De arte alchemica* attributed to Richard de Fournival (Calvet 2018b), and manuals in vernacular languages (French, English, German, and Italian), which are more prolific and contain more detailed descriptions than the often more theoretical texts in Latin.

In the *De arte alchemica*, its author, who expresses himself in the first person, as proof of the personal tone he imprints on his experimentation, hides nothing from us about the transmutatory art, and what tools, materials, and processes are necessary for the work. It is then a real course; but, to our knowledge, the *De arte alchemica* remains an exception in the thirteenth century. The knowledge of the alchemist was not taught, like Aristotle's *Physics* or his *Meteorology*, in the Faculty of Arts, but rather, as we have said, from master to disciple in the alchemist's workshop in the midst of furnaces and the (toxic) fumes of arsenic, sulfur, and mercury. The alchemical art as such was not condemned by the Church, only its fraudulent use (counterfeit money, swindling), or, in the case of the alchemist's failure, his recourse to the invocation of the Devil; the alchemists insisted instead that it should appear as a branch of natural philosophy. One may take as an example the alchemical text called the *Verbum abbreviatum*, which is presented as the work of the Franciscan Roger Bacon, himself having supposedly been trained in alchemy by the Dominican Albert the Great, and who passed on its secret to the Master General of the Order of Friars Minor, Raymond Geoffroy (1289–1295). Similarly, Christine de Pizan mentions the arrival at the court of King Charles V of France, "a cleric living on philosophy and practising alchemy with great talent," a pupil of the pseudo-Arnald of Villanova. The idea of an alchemical school, located in Languedoc and Catalonia, represented by Arnald of Villanova and Ramon Llull, comes from the legends that occur in certain texts where Arnald is shown as instructing Ramon Llull. According to Guglielmo Fabri de Dye, who develops and enriches the legend, both the master and the disciple are summoned to the court of King Edward III of England to inform him of this science so useful for a king who was greedy for gold (Crisciani 2002).

The best way to apprehend the problem of the transmission of alchemical knowledge is to refer to the manuscripts. Indeed, in the fourteenth and fifteenth centuries, unlike in the preceding centuries, the scribes tried to group together texts of the same interest, or those they considered to be the cardinal works of alchemy, those of the pseudo-Albert the Great, the pseudo-Roger Bacon, the pseudo-Thomas Aquinas, and the Franciscan trilogy: the pseudo-Arnald of Villanova, the pseudo-Llull, and John of Rupescissa. Some manuscripts transmit only treatises by a single author, such as the pseudo-Arnald of Villanova or the pseudo-Llull. Moreover, we notice that in several manuscripts from as early as the thirteenth century, the *De mineralibus* of Albert the Great is transcribed at the head, so as to place these writings under the patronage of the Universal Doctor, who in the *De mineralibus* treats alchemy as an important subject to complete the gaps in Aristotle's *Meteorologica* on the subject of metals.

In the fifteenth century, as we have said, texts in the vernacular were disseminated, some of which contained illustrations, especially technical drawings, which, as Geneviève Dumas (2019) reminds us, play an important role in mediating between theory and practice. This alchemical iconography, which over time has improved and become more and more detailed and precise, has an undeniable pedagogical value. During the Renaissance, it found an even more suitable medium for the propagation of alchemical science and its understanding by the public. However, the alchemical texts of the Middle Ages did not yield authority to images. As Jennifer Rampling has explained, they are writings whose authority is never questioned, but on the contrary adapted to the new results that are the outcomes of the experiments carried out by the alchemists (Rampling 2014b). This takes place through a so-called practical exegesis. Indeed, as Rampling notes, each generation of practitioners had to take up the challenge of understanding those who preceded them; witness the numerous annotations and comments in the texts (manuscripts or printed) that try to shed light on what is written in the text they have before them. The aim was to translate authoritative statements into coherent and, above all, reproducible procedures. These conclusions have in turn been included in the textual tradition, with previous sources having been preserved in order to mark a continuity between practical knowledge already acquired and new observations: "In the process, meanings shifted, and terms were reinterpreted in light of textual exegesis and practical experience – a feedback loop that I term practical exegesis" (Rampling 2014b: 21).

To confirm Rampling's point, we can draw on the case of a classic of Latin alchemy, the *Rosarius philosophorum* attributed to Arnald of Villanova. In 1504, Thomas Murchi, the publisher, integrated it into the *Opera omnia* of Arnald of Villanova along with three other alchemical writings, all of which convey the theory of "mercury alone," which was developed at length in the *Rosarius* (Calvet 2006). It seems that, as far as the *Rosarius* is concerned, Murchi took as a model the manuscript Bologna, Biblioteca Universitaria, 104, put together by Johannes de Lachellis de Fontaneto between 1476 and 1477, which is the most important collection of alchemical texts attributed to Arnald. In the last portions of his *Opera omnia*, those dated 1585 and 1586, the *Rosarius* and other treatises were published separately from the medical works. They were then classified as "esoterica," suggesting that the *Rosarius*, as well as other writings of the same type, were the result of a secret teaching by Arnald of Villanova. In reality, this separation, which meant that the master's attested works and the so-called "esoterica" were sold to different clients, was the consequence of a mandate from the Cardinal-Archbishop of Toledo (1584) forbidding the printing of the treatises attributed to Arnald that were related to magic, astrology, or alchemy. The fortune of the *Rosarius* continues through adaptations and commentaries emphasizing, sometimes, the medical virtues of the elixir expressed at the end of the *Rosarius* (e.g. see the Paracelsian Adam of Bodenstein, 1528–1577) and, at other times, the exclusively practical aspects that, according to Andreas Libavius, make this text the typical example of alchemical writing describing a rational process that only a philosopher of nature can decipher.

The problem of interpretation raised by Rampling remains: alchemists of the sixteenth and seventeenth centuries were obliged to base their art, in part, on fourteenth-century texts: the *Rosarius* or the *Testamentum* of the pseudo-Llull. Indeed, the idea that the pseudo-Arnald and the pseudo-Llull were the Pillars of Hercules beyond which there were no longer any undisputed authorities of alchemy became widely accepted in the fifteenth century.

CHAPTER EIGHT

Art and Representation: *The Alchemical Image in the Islamic and Christian Middle Ages*

JENNIFER M. RAMPLING

INTRODUCTION

For readers in late medieval Europe, the alchemical knowledge embedded in diagrams and figurative illustrations held out the promise of both practical instruction and philosophical understanding.[1] Fifteenth-century image cycles depict complex allegories involving human and animal figures, from the iconic "chemical wedding" of Sun and Moon to a parade of allegorical beasts: the red toad drowning in vinegar, the green lion devouring the sun, a pair of serpents pursuing one another's tails. When presented in sequence, these images lay down clues to a tantalizing, interpretative treasure hunt – one that continues to draw in readers even today, as modes of print and digital publication all too often strip alchemical illustrations of their original textual associations and material context.

But what actually makes such images "alchemical?" On the one hand, illustrations encountered in alchemical manuscripts do include features absent from other intellectual or practical traditions, often giving visual expression to the distinctive, image-rich language of alchemical poems and treatises.

This imagery had a prior history – many of the most distinctive features of fifteenth-century Latin and vernacular alchemy are already present in Greek and Arabic writings dating back to the tenth century and earlier, although their signification changed as doctrines evolved and practices varied.

On the other hand, to say that alchemical images are those that illustrate alchemical texts is merely to beg the question. Methods of visualization in alchemical manuscripts are extraordinarily diverse, encompassing not just the charismatic allegories mentioned above, but also geometrical diagrams, sketches of apparatus, and grids and tables packed with text.[2] Image-makers often went beyond their alchemical sources, importing concepts and stylistic details from other scholarly disciplines or juxtaposing alchemical ideas with religious iconography. Both the matter and the form of alchemical imagery were shaped by changing cultural contexts, adjusted to incorporate references from scripture, classical verse, and romance literature, as well as tenets of Scholastic natural philosophy and Neoplatonic cosmology. We might ask at what point, then, an image ceases to belong to another tradition and becomes "alchemical" – or vice versa.

There is, as yet, no definitive typology of alchemical imagery, although scholars have suggested several useful working distinctions, while emphasizing the need to treat such categories flexibly. Thus Barbara Obrist distinguishes between nonverbal methods of visualization and those that incorporate verbal (or textual) elements, including tables and lists (2003). Mino Gabriele proposes a threefold classification: technical drawings of laboratory apparatus; codes, ciphers, and geometrical combinations (often involving alphabets, sometimes with mnemonic or magical connotations); and figurative imagery (1997: 29–33).

For the purposes of this chapter, I propose a further typology based on the *function* of images, as well as the form – a classification that allows us to distinguish between "alchemical" and "nonalchemical" usage. How, for instance, should we categorize the depiction of a glass distillation vessel (Figure 8.1)? Depending on context, it may be read as alchemical, but our glass might also illustrate the preparation of mineral acids in a goldsmith's workshop or the distillation of pharmaceutical remedies in an apothecary's. Our hypothetical vessel might be sketched as an instructional diagram in the margins of a recipe book, adapted as the frame for a complex allegory taking place within the flask, or positioned in the background of a generic scene satirizing false alchemists. It may even fulfill several functions within the same image.

Of my three broad, functional categories, the first – and by far the largest – comprises technical drawings of laboratory furnishings, including vessels, furnaces, and apparatus. Both Arabic and Latin manuscripts frequently include such figures, which typically offer visual support for practices outlined in the main text. They can vary enormously in size, style, and complexity, from small diagrams, inset into text or jotted in margins, to full-page, intricately labeled schemata. While often encountered in alchemical contexts, they may also appear in other practical fields where chemical apparatus is employed, such as metallurgy and medicine.[3]

FIGURE 8.1 Allegorical figure within the alchemists' flask. Detail of Ripley Scroll, Beinecke Rare Books & Manuscript Library, MS Mellon 41. By permission of the Beinecke Rare Books & Manuscripts Library.

The second category includes those images that we might think of as properly *alchemical*: that is to say, figures devised by proponents of alchemy to evoke or elucidate various aspects of the art. This category offers even richer diversity than the first, encompassing everything from linear, diagrammatic forms to naturalistic, figurative depictions. As we shall see, such images also occupy a wide spectrum in terms of function (intended or actual) – whether expressing theoretical concepts, allegorizing practical processes, or serving to elevate and disguise "secret" knowledge.

The third and final category consists not of alchemical images so much as images *of alchemy*: namely, those that describe or allude to alchemical practice but are not necessarily produced by alchemists or intended to expound the content of their art. Portraits of alchemists fall under this head, as do satirical woodcuts and other popular representations of alchemical fraud. Such depictions may tell us a great deal about the reception and reputation of alchemy in different times and places, including its role in popular culture, but they are less reliable as sources of insight into alchemy as actually theorized and practiced.

While the borders between categories are highly permeable, this chapter will focus on the second, namely on images that shed light on alchemy as conceptualized by its practitioners during the Islamicate and Christian Middle Ages. Asking what makes such images alchemical also allows us to reflect on the kind of work alchemical images do – in particular, what work they do that cannot be served by text alone. Images may extend or complement text in various ways: thus diagrams can serve as tools for classifying information and visualizing relationships, while figurative imagery creates opportunities to dramatize otherwise obscure chemical properties and procedures as allegorical narratives. Both diagrammatic and figurative imagery may express analogies with other branches of knowledge, often with a view to elevating the prestige of alchemy as a philosophical and divinely sanctioned science. By drawing on the style and content of figures developed in other intellectual disciplines, image-makers could illustrate – quite literally – alchemy's relationship to diverse authorities, methods of reasoning, and sources of inquiry, even in cases where these relationships are not alluded to directly in the text.

While images convey more than text alone, this does not mean we should study them in isolation from written sources, even in cases where the image dominates, as in the case of one of the most spectacular emblematic productions of the late fifteenth century, the so-called "Ripley Scroll" (discussed further below). Knowledge of textual traditions provides necessary insight into the principles that may underlie a particular design while also revealing points of divergence in theory and practice. Alchemical imagery is complex, syncretic, and adaptive, and we can learn a great deal from cases where an artist or designer departs from the letter of the text to generate new, creative readings of past authorities.

VISUALIZATION USING DIAGRAMS

Alchemical imagery did not gradually develop from primitive, early figures, but was already allegorical, figurative, and ripe for adaptation from the start of the tradition. Figurative illustrations are nonetheless in the minority when we take all categories of alchemical imagery into account, especially drawings of apparatus, which greatly outnumber figurative representations in both Arabic and Latin manuscripts. Rather than being superseded, however, these simpler methods of visualization coexisted alongside more elaborate forms while serving different functions.

Graphic methods of verbal organization, including lists and tables, appear in some of the earliest works of Arabic alchemy. For instance, the Jābirian "science of the balance" uses tabulation to apportion numerical values to the letters that make up the names of metals – as in the *Kitāb al-aḥjār ʿalā raʾy Balīnās* (*The Book of Stones According to Balīnās's Opinion*), where tables present this information concisely and clearly.[4] Such "verbal diagrams" need not be intrinsic to a text: used as an annotation technique, they also reveal the efforts of readers to classify and epitomize the content of their reading. Scribbled in the margins of manuscripts or added to the end of texts, simple diagrams served to group together related terms or to offer brief taxonomies of names and properties (Figures 8.2 and 8.3).

Some of the earliest alchemical diagrams also borrow from visualization techniques developed in other branches of scientific or philosophical inquiry,

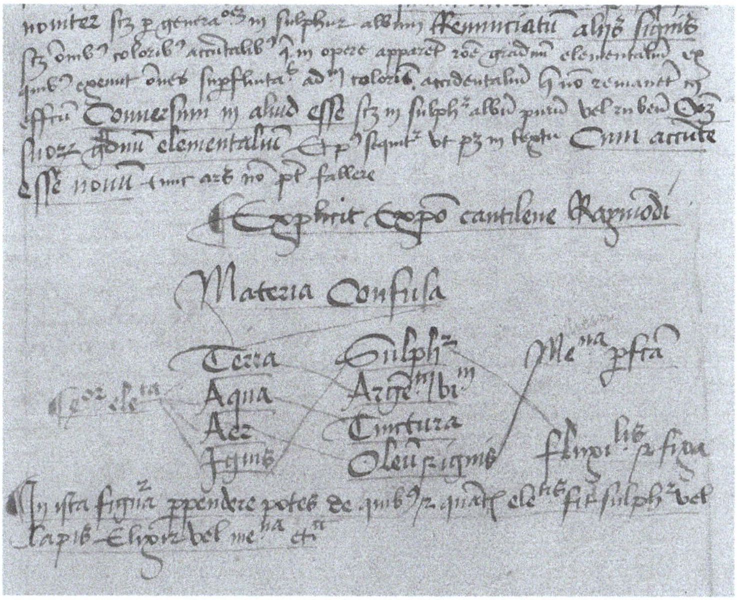

FIGURE 8.2 Giles Du Wes, exposition of the pseudo-Llullian *Cantilena* (composed ca. 1332), added to the mid-fifteenth-century manuscript ca. 1506. Beinecke Rare Books & Manuscript Library, MS Mellon 12, fol. 163v. By permission of the Beinecke Rare Books & Manuscripts Library.

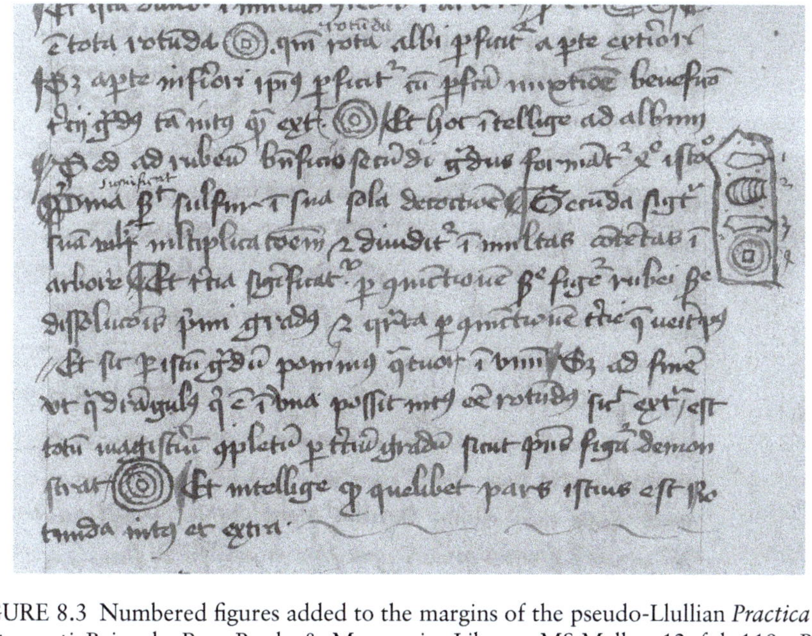

FIGURE 8.3 Numbered figures added to the margins of the pseudo-Llullian *Practica Testamenti*. Beinecke Rare Books & Manuscript Library, MS Mellon 12, fol. 119r. By permission of the Beinecke Rare Books & Manuscripts Library.

sometimes adapted to include figurative elements. By adopting conventions from other disciplines, writers in both the Arabic and Latin traditions also sought to position alchemical content within wider cosmological schemes or to show how it related to different intellectual traditions. The most influential Islamic alchemist of the fourteenth century, ʿIzz al-Dīn Aydemir al-Jildakī, seems to have adapted astronomical diagrams as the basis for the intricate figures in the *Kitāb al-Burhān fī ʿilm asrār al-mīzān* (*The Book of Demonstration: On the Knowledge of the Secrets of the Balance*), a work that expounds the cosmological foundations of alchemy.[5] Constantine of Pisa, active in Italy during the 1250s, drew on astrological charts and cosmological figures in his own treatise (Obrist 1993).

Such intellectual borrowings had particular urgency in thirteenth-century Europe, where alchemy was a newcomer to the arts, lacking classical precedent – putting the onus on alchemy's adherents to justify its value as a branch of learned knowledge rather than a purely manual practice. While one solution was to frame alchemy in terms of the principles of Aristotelian natural philosophy, Scholastic writing generally did not encourage metaphorical language or pictorial representation. Unsurprisingly, the alchemical images that accompany such treatises tend to be diagrammatic rather than figurative.

The set of diagrams that accompanies Constantine of Pisa's *Liber secretorum alchimie* (*Book of the Secrets of Alchemy*, 1257) is one of the earliest examples of alchemical visualization in the Latin tradition. Here, the use of diagrams suggests an attempt to reconcile natural philosophy with the biblical account of creation; deferring to Aristotle while also admitting elements of Platonic cosmology (Obrist 1993: 117). Constantine himself seems to have viewed his tables and diagrams as having pedagogical value, explaining that, "since these things transcend human understanding, it is necessary to [help] the intellect recall them through visible likeness" (Obrist 1990: 101). In practice, his "likenesses" draw from existing pictorial traditions. The most striking, a creation diagram made up of overlapping, semicircular segments, maps the creation of the six metals onto the six days of Creation – the seventh metal, mercury, providing the chaotic prime matter at the start of the work. As Barbara Obrist suggests, these diagrams offer a form of "visual substitution" by adapting earlier pictorial forms to accommodate new content: in this case, a theory of metallic generation (Obrist 1993).

Other early diagrams borrow more conventionally from Aristotelian philosophy. The square of opposition, a logical diagram used to illustrate the relationship between Aristotle's primary qualities (hot, cold, wet, and dry), is obviously relevant to those theorizing material transformations. Versions of the square appear in several alchemical treatises in manuscript, including the early fourteenth-century *Ysocedron* of the Benedictine Walter Odington, also known as Walter of Evesham (fl. ca. 1280–1301; MS Digby 119, fol. 147r; cf. Thorndike 1923–1958: 4:127–32; Thomas 1968). The square appears in some copies of an even more influential fourteenth-century treatise, the *Rosarius philosophorum* (*Rosary of the Philosophers*; incipit "Desiderabile desiderium"), attributed to the English alchemist John Dastin, although here it is unclear whether the figure is integral to the text or a later addition (Figure 8.4).

Other diagrams borrowed from logic are more specific in purpose: most famously, those employed in the large body of alchemical writings pseudonymously attributed to the Majorcan philosopher Ramon Llull, or Raymond Lull (ca. 1232–1316). The historical Llull developed a sophisticated approach to figures in his Art: a logical system that incorporated philosophical terms into a variety of pictorial forms, including wheels, ladders, trees, and tables – diagrams that in turn reveal the influence of theological *figurae* crafted by the Franciscan visionary Joachim da Fiore (Reeves and Hirsch-Reich 1972). Llull's alchemical followers would adopt many of the same forms, although their diagrams do not necessarily operate in the same way. Whereas Llull's combinatorial art uses concentric wheels to combine terms and generate syllogisms, pseudo-Llullian texts employ diagrams more straightforwardly to express relationships between chemical materials and processes.

The foundational text in the pseudo-Llullian tradition, the *Testamentum* (ca. 1332), uses triangular and circular figures to illustrate both theoretical and

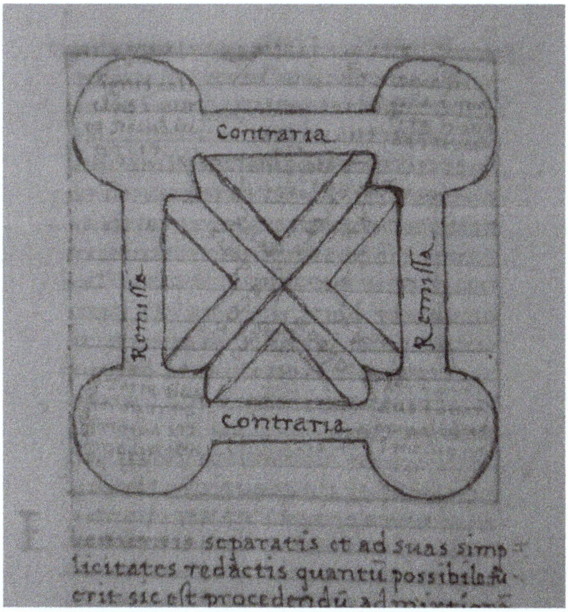

FIGURE 8.4 Aristotelian square of opposition in the *Rosarius philosophorum* (incipit "Desiderabile desiderium impreciabile precium") attributed to John Dastin. Beinecke Rare Books & Manuscript Library, MS Mellon 28 (ca. 1525), fol. 4r. By permission of the Beinecke Rare Books & Manuscripts Library.

practical doctrines, from the genesis of metals to the ingredients required in specific processes (Pereira and Spaggiari 1999). In the first book, the *Theorica*, the writer describes how metals are generated in successive stages from their primordial constituents, recording each stage on a simple wheel, possibly as a mnemonic device. In the *Practica*, figures are used to plot the combination of ingredients into successively more complex compounds (Figure 8.5).

For instance, a triangular figure represents the combination of three ingredients: B (mercury), C (saltpeter), and D ("vitriol *azoqueus*," a substance that may not correspond to normal vitriol, a metal sulfate). Inside the triangle is a circle labeled E, denoting the product of BCD: a solvent identified in the text as the "Stinking Menstruum." The same menstruum, E, then appears as a component of the next two triads, as a solvent used to dissolve silver (F) and gold (H), respectively.

While transmutation is the end goal of this particular operation, elsewhere in the pseudo-Llullian corpus figures denote the organic ingredients used in alchemical medicine – in particular, the extraction of "quintessences" using the distilled spirit of wine, a practice adapted from John of Rupescissa. The writer of the *Liber de secretis naturae, seu quinta essentia* (*Book Concerning the Secrets of Nature, or the Quintessence*) incorporated a variety of substances, including honey and red wine, into his own alchemico-medical practice, once more using a wheel to plot these ingredients (Figure 8.7). While these figures typically consist of lines and letters only, some combine figurative and tabular elements, like the spectacular "tree" that accompanies several early copies of the *Liber de secretis naturae*, including a fifteenth-century example now held in the Beinecke Library (Figure 8.6).

In these core pseudo-Llullian writings, the placement and composition of figures is integral to the writers' vision of the work as a whole. Studied in conjunction with the text (which usually includes detailed exposition of the alphabets used in the figures), these diagrams functioned as tools, allowing readers to proceed from theoretical, general principles to the more particular kinds of knowledge required to actually make the stone. For the writer of the *Testamentum*, images were vital elements of this process: by diligently studying them, students of alchemy might grasp not only which ingredients were required, but also *why* a particular combination of materials must necessarily create a given substance. As the writer warns, "Unless you know the said alphabet by heart, you cannot practice, nor can you even begin" (Pereira and Spaggiari 1999: 314; Rampling 2020: 104).

In this respect, diagrammatic figures helped to shore up the prestige of the text as a "philosophical" treatise in the tradition of a respected authority; in this case, Llull. However, it is not clear that all readers valued the figures equally. While some copyists faithfully reproduced the elaborate diagrams, other copies of the *Testamentum* relegated them to space in the margins, omitted them entirely, or grouped them in one place, usually toward the end of the text (Hinckley 2017). For those interested primarily in gleaning practical and doctrinal information, the complex system of figures may have seemed distracting or immaterial to the task at hand.

FIGURE 8.5 Figures describing a succession of compounds in the pseudo-Llullian *Practica Testamenti*. Beinecke Rare Books & Manuscript Library, MS Mellon 12, fol. 99v. By permission of the Beinecke Rare Books & Manuscripts Library.

The opposite situation could also arise, with readers inserting images into text in order to make sense of written instructions. For instance, the writer of the "Desiderabile desiderium" attributed to John Dastin used a geometrical argument in order to explain one of the most vexing problems in medieval alchemy – how to determine the correct "proportion" of ingredients to use in the stone. Rather than squaring the circle, practitioners should learn how to "circle the square" by transforming the "quadrangle" of the four elements into a single, perfectly proportioned substance, signified by a circle. Using a compass, one should first draw a circle, then place a square within it, such that

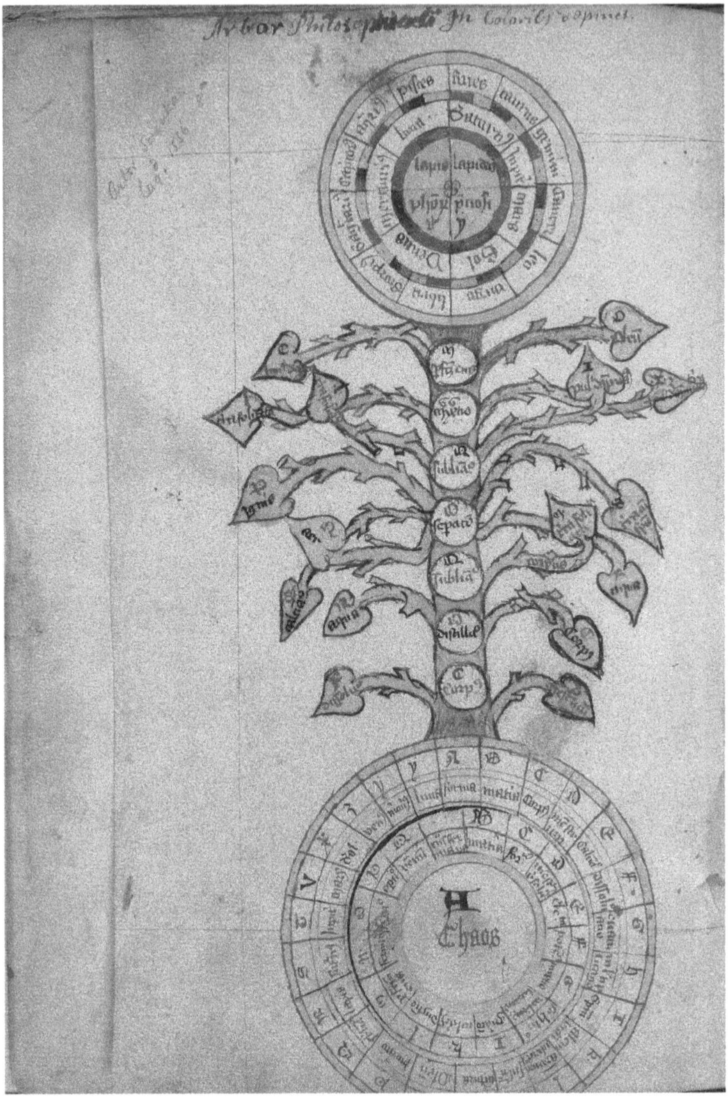

FIGURE 8.6 Alchemical tree in the pseudo-Llullian *Liber de secretis naturae seu quinta essentia*. Beinecke Rare Books & Manuscript Library, MS Mellon 12, fol. 277v. By permission of the Beinecke Rare Books & Manuscripts Library.

each corner touches the circumference of the circle. The square is divided into twelve equal parts by drawing lines from the circle's center to the corners of the square, then dividing each of the four larger triangles into three. This results in twelve equally proportioned triangles, which can be used to plot a circle. "Therefore by twelve triangles," says Dastin, "the quadrangle becomes round" (Dastin 1702: 2:318). While it is unclear whether a diagram was included in the

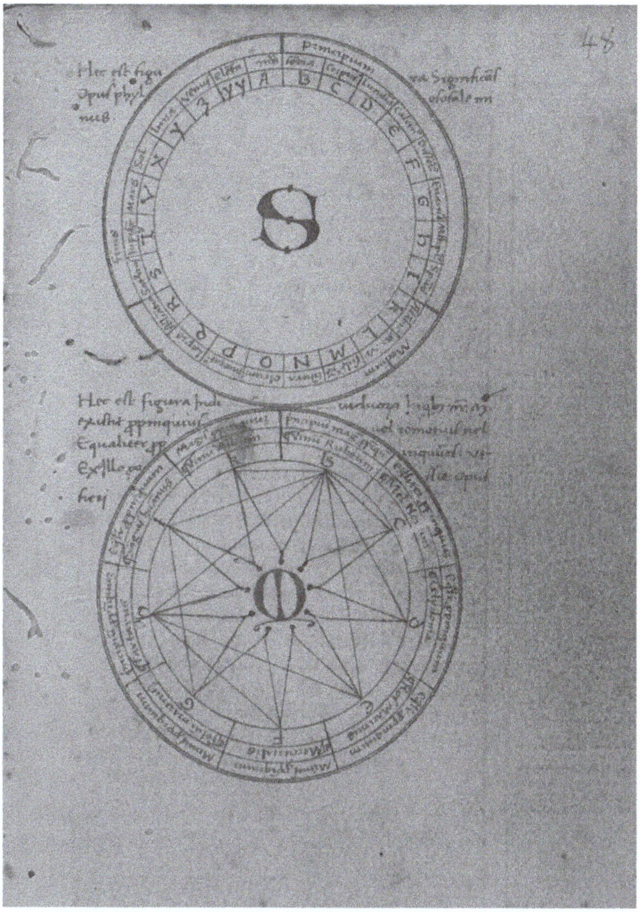

FIGURE 8.7 Circular figures in the pseudo-Llullian *Liber de secretis naturae seu quinta essentia*. Science History Institute, Othmer Library MS 7 (1498). By permission of the Science History Institute.

original text, several later readers added figures of their own, suggesting that they found a drawing helpful as a means of visualizing the figure described in the text (Figure 8.8).

Yet, although we might expect images to increase the accessibility of alchemical doctrines and widen the readership of any accompanying text, the evidence is ambiguous. For instance, where diagrams are developed from existing theoretical systems, it is more likely that a limited, learned audience is intended. As the popularity of pseudo-Llullian techniques gathered pace among lay audiences during the fifteenth century, it is not obvious how the figures were read, or what practical role (if any) they played, although commentators on the alchemy of "Raymond" generally seem to have been more concerned with its practical underpinning than

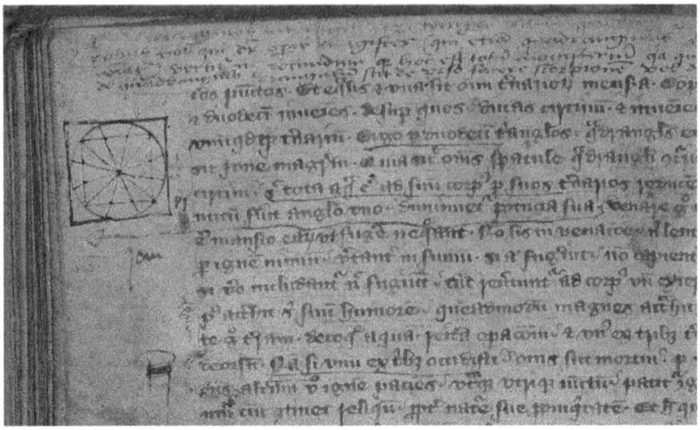

FIGURE 8.8 A later reader's attempt to sketch the geometrical figure described in a copy of the *Rosarius philosophorum* attributed to John Dastin. Cambridge, Trinity College Library MS O.2.18 (late fourteenth/early fifteenth centuries), fol. 88v. By permission of the Master and Fellows of Trinity College, Cambridge.

with reproducing Llullian-style diagrams. While the Italian alchemist Christopher of Paris incorporated pseudo-Llullian alphabets into his own treatises, these are not accompanied by figures. The English canon George Ripley (fl. 1470s), who played an important role in popularizing pseudo-Llullian doctrines in England, dispensed with sequential diagrams in favor of a single "master" diagram in his Middle English poem, the *Compound of Alchemy* (1471). This figure, which Ripley describes as his "Wheel" and the key to his poem, is unusual in combining Llullian-style concentric circles with allegorical verse. Here, the original philosophical context of Llullian diagrams has been stripped away to leave forms familiar from other medieval diagrammatic traditions: quaternities of elements, bodily humors, seasons, compass points, and the planetary spheres of the Aristotelian cosmos. The diagram is further embellished with English allegorical verses that recount the sojourn of "the red man and his white wife" through purgatory and into heaven. In crafting his own diagram, Ripley thus introduces figurative imagery to create a hybrid form; one that seems to have appealed to readers, to judge by the relatively large number of complete and partial copies that survive in manuscript (Rampling 2013).

FIGURATIVE IMAGES: EARLY EVIDENCE

When tracing the development of alchemical imagery, we might expect a gradual evolution in style and complexity, from schematic diagrams to more elaborate, figurative depictions. In fact, allegorical language is already present in Greek and Arabic texts from an early date, including many figures and tropes that would later be adopted into Latin works. The development of alchemical image-making cannot be separated from this textual tradition. Medieval alchemical images are often closely related to texts and frequently evoke the same metaphors and analogies as those encountered in alchemical writings, including works produced at other times and in other languages. This image-rich metaphorical language appears to be integral to the tradition of alchemy generally and cannot simply be attached to one historical period.

The long tradition of Islamicate alchemical imagery and its Latin reception has received little in the way of systematic study, partly due to the challenges of working across materials written in Greek, Arabic, and Latin. Contrary to popular belief, there is a rich tradition of alchemical depiction in the Arabic tradition, dating back to at least the tenth century and possibly associated with Greek exemplars that are no longer extant (Hallum 2009; cf. Haldane 1978; Contadini 2007). While the small number of examples of Arabic alchemical illustration makes it difficult to propose typologies, that number also reveals considerable variety.

Although no tenth-century illustrations survive to bear witness to the style of alchemical imagery circulating at this early date, several suggestive descriptions are found in texts of the period. *Rumūz*, the obscure terms used by alchemists to disguise concepts and practices, are discussed in the *Rutbat al-ḥakīm* (*Scale of the Sage*), written by Maslama b. Qāsim al-Qurṭubī in al-Andalus in the first half of the tenth century (Fierro 1996). Maslama seems to have viewed *rumūz* as encompassing figurative representations, since he refers explicitly to figurative symbols when discussing the Birbā temple in Ikhmīm (Panopolis), asserting that these images are alchemical in nature (Callataÿ and Moureau forthcoming). Another telling account appears in one of the most influential Arabic alchemical texts, the *Mā' al-waraqī wa al-arḍ al-najmiyya* (*The Silvery Water and the Starry Earth*, later translated into Latin as the *Tabula chemica*) of the tenth-century writer Ibn Umayl, which describes his discovery of an illustrated book in an underground vault (*Theatrum chemicum* 1602–1661: 5: 193–4; Stapleton 1933).

Ibn Umayl adapted his account from a still earlier work, the *Kitāb sirr al-khalīqa* (*Book of the Secret of Creation*) falsely attributed to Apollonius of Tyana (Balīnūs in Arabic). Balīnūs describes digging beneath a statue of Hermes to discover an underground chamber. Within, he finds an old man sitting upon a golden throne, clasping a text inscribed on an emerald tablet and a book of secret wisdom, which Balīnūs takes for himself. In retelling this story, Ibn Umayl relocated Hermes' tomb to the interior of an Egyptian pyramid and also made the content more explicitly alchemical. Of particular note, he conflated the emerald tablet and the book of wisdom into a single, stone book inscribed with symbolic images. These include the figure of two birds – one winged, one wingless – chasing one another's tails, as well as images of the sun and crescent moon. Ibn Umayl explicates the figures and their meaning in a lengthy description and commentary, compounding the value of alchemical symbols as a means of conveying secret knowledge in a compressed yet authoritative form. The scene would later be illustrated in Arabic manuscripts (Berlekamp 2003) as well as in the Latin tradition.

While texts like *The Silvery Water* hint at an early and prestigious role for figurative alchemical imagery, the oldest surviving manuscripts were produced several centuries later. One of the earliest examples is the *Muṣḥaf al-ṣuwar* (*Tome of Images*), a work in thirteen books presented as a dialogue between the third/fourth-century CE authority Zosimos of Panopolis and his pupil and interlocutor, Theosebeia. As Benjamin Hallum has convincingly argued, the text of the *Tome of Images* is not an original Greek work, but more likely a product of the tenth century, compiled from Arabic translations of earlier Greek writings, including authentic works of Zosimos (Hallum 2008; Hallum 2009). The oldest manuscript, made in Mamluk Egypt in 1270 and now in Istanbul, includes a dazzling sequence of painted allegorical images.[6] Importantly, the illustrations supply information that cannot be obtained from the text alone. For instance, Zosimos and Theosebeia are depicted several times, but their figures seem to be more than portraits. Zosimos is usually drawn accompanied by the sun (implying hot and dry qualities) and Theosebeia by the moon (cold and moist), suggesting that they here stand in for opposing male and female principles in the alchemical work (Hallum 2008: 269). Similar imagery would develop later in the Latin tradition.

Another illustrated work produced in Mamluk Egypt is the *Kitāb al-aqālīm al-sab'a* (*Book of the Seven Climes*) written in the mid-thirteenth century by Abū al-Qāsim Muḥammad ibn Aḥmad al-'Irāqī. In what was already a common trope in alchemical writing, al-'Irāqī claims to have drawn the book's contents from other, more ancient sources, including a secret book of Hermes Trismegistus. Not all of his sources are so temporally distant, however, since at least three of the book's illustrations are adapted from the *Tome of Images* – possibly even from the Istanbul manuscript. These include an image of a human body held aloft by three figures, to which al-'Irāqī has added wings (in the *Tome*, they are wingless).[7] Other illustrations include pseudo-hieroglyphics, as well as images apparently copied from authentic Egyptian monuments (Hallum and Marée 2016). In this respect, the images contribute to a long-lived tradition, encountered in other Arabic sources, which sought to place the origins of alchemy in Pharaonic Egypt (Martelli 2017: 330–1). Unfortunately, al-'Irāqī's text does not offer much help in understanding the nature of the images themselves, which remain enigmatic.

ALLEGORICAL IMAGES: THE LATIN TRADITION

Most Islamicate illustrations did not survive the transition from Arabic to Latin during the great translation movement of the twelfth and thirteenth centuries. However, the image-rich language of translated texts, including works like the *Tabula chemica* and *Turba philosophorum* (*Crowd of Philosophers*), furnished ample material for subsequent illustrations. Many of the Arabic cover names, or *Decknamen*, used to disguise the identity of ingredients lent themselves readily to visual representation, including the use of a serpent or dragon to signify mercury, or the representation of vitriol as a green lion – images that, as we shall see, later became staples of European image cycles.

Perhaps the most evocative image in these newly translated texts is the "chemical wedding" of sun and moon. Arabic alchemical writings frequently describe the "marriage" of principles or ingredients, a relationship that we have already seen alluded to in the *Silvery Water* and illustrated in the *Tome of Images*. Ibn Umayl elaborated this allegory further in his alchemical poem, the *Risālat al-Shams ilā al-Hilāl* (*Letter of the Sun to the Crescent Moon*), framed as a dialogue between sun and moon. The Latin version of the poem, the *Epistola solis ad lunam crescentem* (the author's name morphing into "Senior Zadith"), inspired both writers and image-makers, as did Ibn Umayl's account of the illustrated emerald tablet (Figure 8.9).

Another source for the chemical wedding, the *Visio Arislei* (*Vision of Arisleus*), was probably translated from the *Risālat madd al-baḥr dhāt al-ru'yā* (*Epistle of the Rise of the Sea, Containing a Vision*; Moureau 2020: 121). This allegorical text describes the incestuous union of the prince Gabricus and his sister Beya, signifying the alchemical principles of sulfur and mercury, respectively.

One of the earliest visual representations of the chemical wedding was made to accompany not Latin treatises, but vernacular verse: an untitled, Middle Dutch poem written in the second half of the fourteenth century by an author who identifies himself as Gratheus (Birkhan 1992). In this poem, partly derived from translations of Latin sources, Gratheus offers an introduction to alchemical ideas and practices for non-Latinate readers, asserting that he writes in Dutch "for the benefit of all those Ladies and Gentlemen who do not understand Latin" (Birkhan 1992: 1:191–5; Pereira 1999: 348). The accessibility of the work was surely aided by its combination of verse, image, and mnemonic phrases, including idiosyncratic labels for ingredients and apparatus (Obrist 2003).

Gratheus's sequence of small, "thumbnail" illustrations includes the earliest example of what would eventually become a common device – the use of human figures to represent chemical processes taking place within the frame of the alchemical flask. Inside the vessel, Gratheus's characters act out a drama already familiar from Arabic and Latin texts: the wedding of a king and queen who represent opposing principles (here named Ylarius and Virgo) and the generation of their philosophical child. Both textual and visual imagery reveal the influence of Ibn Umayl, and it is surely no coincidence that the same manuscript includes a Middle Dutch translation of the *Tabula chemica* (Marinovic-Vogg 1990). Such images could also be added to texts at a later date, as in the case of the German "picture-poem" (*Bildgedicht*) *Sol und Luna*, probably composed early in the fifteenth century, which once more reveals the influence of the *Tabula chemica* and *Visio Arislei* in its recounting of the courtship of Gabricus and Beya. During the sixteenth century, the German poem was illustrated with a sequence of figures charting their nuptials, coitus, death, and dissolution into a single, dual-headed, hermaphroditic body.[8]

FIGURE 8.9 The Emerald Tablet, *Aurora consurgens*. Zurich Zentralbibliothek MS Rh. 172 (fifteenth century), fol. 3r. By permission of Zurich Zentralbibliothek.

Despite its non-Christian origins, the image of the chemical wedding acquired a further level of allegorization through the incorporation of religious themes. In fact, one of the most distinctive and remarkable features of European alchemical imagery is its fusion of Arabic alchemical tropes with Christian iconography. In both texts and images, the passion, death, and resurrection of Christ offer a powerful analogy for the tribulations of mercury, which must endure calcination, sublimation, and putrefaction on its way to ultimate perfection (Gabriele 1997: 71–2). Indeed, comparisons between the perfected matter of the philosophers' stone and the resurrected Christ seem to have been irresistible for image-makers in Christian Europe. In the illustrations to Gratheus's poem, for instance, one glass vessel includes the head of Christ, here signifying the stone (Birkhan 1992: 2:86; Obrist 2003).

For alchemical authors, the use of figurative language in biblical prophecy and parable already suggested a powerful exemplar and analogue for "philosophical" speech. Theologians were used to reading scripture on multiple levels, providing allegorical interpretations of literal accounts, and drawing analogies between the events of the Old and New Testaments (De Lubac 1959–64). While a bare, literal reading was insufficient to reveal the full sense of scripture, the ability to uncover hidden layers of meaning was a gift not granted to everyone – suggesting that such exegetical insight was divine in origin.

As with other types of alchemical imagery, Christian allegories were already worked out in written sources before manifesting as visual content. The notion that alchemical knowledge was a gift of God (*donum dei*) was reinforced in fourteenth-century treatises, including the *Pretiosa margarita novella* (*Precious New Pearl*) of Petrus Bonus of Ferrara and the corpus of writings pseudonymously attributed to Arnald of Villanova. Indeed, one pseudo-Arnaldian tract, the *Tractatus parabolicus*, is entirely devoted to working out an analogy between Christ and mercury. Pseudo-Arnald notes that, since Christ is the exemplar for all things, his incarnation can shed light on the art of the philosophers: "according to Christ's conception, generation, nativity, and passion, it is possible to come to the knowledge of our elixir by comparing it to Christ, and as foretold by the prophets" (Calvet 2011: 532).

While the Christ figure is only briefly referenced by Gratheus, the melding of alchemical and religious iconography would play a more prominent role in shaping one of the most famous alchemical image cycles, *Das Buch der heiligen Dreifaltigkeit* (*The Book of the Holy Trinity*). This German treatise was produced around 1416 at the time of the Council of Constance, and the anonymous author writes that he presented a summary of the work to the Holy Roman Emperor, Sigismund, who was attending the Council – a fact reflected by the extensive use of imperial imagery throughout the illuminations, including heraldic eagles (Obrist 1982).

In the *Holy Trinity*, as in the *Tractatus parabolicus*, the passion and death of Christ are related to chemical practices. Throughout, the author uses color coding to refer to the metals, although this coding is not applied consistently and would probably be impossible to decipher without the guidance of the text. For instance, the calcination of mercury is allegorized as a series of execution scenes in which mercury containing "Mars" (iron) is represented by Christ crucified beneath a gallows, while mercury containing "Venus" (copper) is instead decapitated (Obrist 1982: 150–1). In neither case, however, are colors used to distinguish between the different metals linked to mercury (Figure 8.10).

FIGURE 8.10 The coronation of the Virgin, *Das Buch der heiligen Dreifaltigkeit*. John Rylands Library, MS German 1 (fifteenth century), fol. 9r. Copyright of The University of Manchester.

On the other hand, color is used to correlate various elements of an elaborate scene depicting the coronation of the Virgin Mary – a popular subject in medieval religious iconography, here recast as the combination of metals and elements (Figure 8.11).

Mars and the element of fire are both represented by the color red, and further identified with the figures of God the Father and the Evangelist Luke. Christ and the angel signifying Matthew, both robed in green, evoke Venus (and the element of water), while John, shown as a black eagle, indicates Saturn (lead and the element of earth; Obrist 1982: 165–7). Throughout, the designer of the image sequence repurposes Christian imagery to generate an extended allegory on the properties of the stone, with particular weight given to the relationship between Jesus Christ and his mother, the Virgin Mary. The analogy culminates

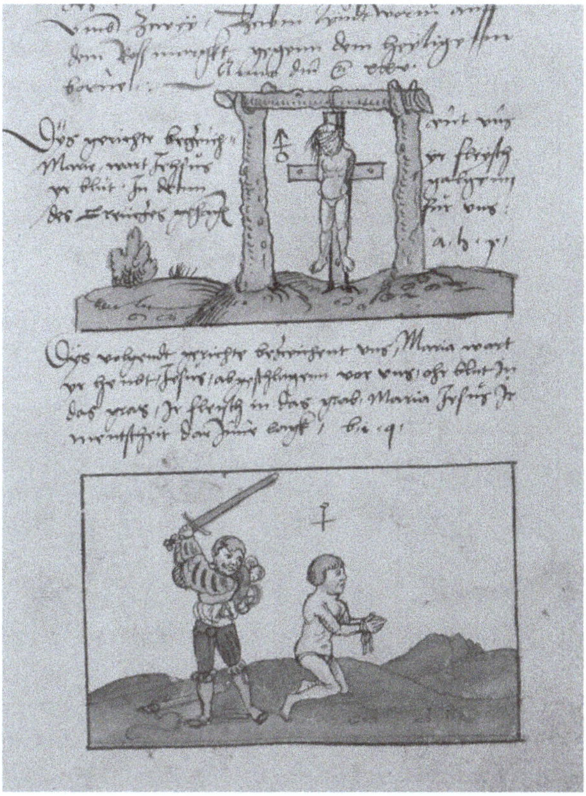

FIGURE 8.11 Calcination of mercury, *Das Buch der heiligen Dreifaltigkeit*. Leiden University Library, MS Vossianus chym. 29 (sixteenth century), fol. 76r.

in a play on the well-known tropes of the chemical wedding, as Christ and Mary are fused into a single, alchemical hermaphrodite: male and female principles harmoniously united in the form of the stone.

In a curious case of iconographic cross-pollination, much of the imagery of the *Holy Trinity* repeats in the second of the great, early fifteenth-century image cycles, *Aurora consurgens* (*The Rising Dawn*). This lengthy Latin treatise, written around 1420, is pseudonymously attributed to Thomas Aquinas, and divided into two books, only the second of which is illustrated (Obrist 1982; Gabriele 1997; Crisciani and Pereira 2008). Both the text and the series of illustrations provide points of connection with the medieval Arabic alchemical tradition by offering a commentary on the ubiquitous *Tabula chemica* of Senior Zadith (Ibn Umayl). One striking image reconstructs Senior's description of the tomb of Hermes, including the seated figure of the ancient sage holding open a book. As in the Arabic tradition, the book is represented not as a text, but as an illustrated tome, although the vault has been transformed from a pyramid into a Christian church. It appears that the allusive images inscribed on the emerald tablet of Hermes Trismegistus, the ancient patriarch of alchemy, offered an important precedent for image-makers in both the Arabo-Muslim and Latin Middle Ages as a venerable authority for their own illustrative traditions.

REPRESENTING PROPERTIES AND PROCESSES

What, however, did allegorical images have to offer readers that could not be expressed by text alone – or by other forms of visualization? While both diagrams and figurative elements can serve to distill complex information into a single image, the latter are generally less suited to conveying detailed, technical, or quantitative information. We need only think of the precision of pseudo-Llullian diagrams, where letters provide clear links to alphabets and definitions discussed in the anchor text, or a cosmological diagram, like Ripley's *Wheel*, which records the proportion between metals used in making the stone (Rampling 2013: 58–9). On the other hand, ease of interpretation is generally not the remit of figurative alchemical imagery, which tends to obscure rather than clarify philosophical speech.

Allegory can nonetheless convey qualitative meanings that are less readily expressed through diagrams: in particular, the distinctive properties of substances and the dynamic nature of chemical processes. Visual metaphors readily evoke the chemical properties of substances that are not apparent in more naturalistic depictions, but which still can be identified by viewers knowledgeable about both materials and the textual tradition that records their operations. Birds and feathers might suggest volatility, for instance, while an earth-dwelling creature, such as a toad, implies fixity or corruption. Solvents are typically represented as carnivorous beasts, from the green lion to the mercurial dragon.

Above all, alchemical imagery strove to capture the peculiar physical and chemical properties of mercury (Lüthy and Smets 2009: 411–20). Already in Greek and Arabic sources, mercury's special properties are alluded to using a host of cover names. Its fluidity and volatility – not to mention the penetrative, amalgamating power that enables it to swiftly "devour" gold – opened up a host of possible depictions, whether as a flowing fountain, bird in flight, or venomous serpent. Multiple identities could be expressed in a single image: thus the *Aurora consurgens* sequence includes a figure that alludes to mercury's triple nature as fire, soul, and the ashes of a serpent. These mercurial identities, garnered from Arabic sources including the *Turba philosophorum* and *Tabula chemica*, are visually translated into a flock of small, infant figures (souls) fleeing a fire that has already consumed the serpent's body (Gabriele 1997: 77) (Figure 8.12).

The properties of mercurial compounds also merited special representation. In the *Aurora*, for instance, a deadly basilisk (drawn in a style familiar from contemporary bestiaries) may symbolize the red, toxic cinnabar from which mercury is extracted; a mineral that Pliny compared to the basilisk's blood (Gabriele 1997: 75) (Figure 8.13).

FIGURE 8.12 Threefold mercury, *Aurora consurgens*. Leiden University Library, MS Vossianus chym. 29, fol. 76r.

FIGURE 8.13 The basilisk repelled by a mirror, *Aurora consurgens*. Zurich Zentralbibliothek MS Rh. 172, fol. 41v. By permission of Zurich Zentralbibliothek.

Since these interpretations allude to mercury's known properties, they operate differently from the Christ–mercury parallel discussed above. There, the analogy is forged not with a specific substance, but with the succession of procedures that mercury is subjected to, resulting in an overarching narrative of mercury's trials and resurrection. The sequential nature of such processes results in changes to the properties of a substance, a transformation that is most obvious in the case of color. The traditional progression of color changes in Latin alchemical imagery is from *nigredo* (the blackening or putrefaction of the prime matter) to *albedo* (the whitening of the stone) to *rubedo* (reddening). The *nigredo* might be represented as a crow or raven, to indicate blackness, or as a dead body to invoke the death or mortification of matter. The white and red stones can also be represented as animals, birds, or human figures, as in *Aurora consurgens*, where three birds – black, white, and red – are shown wrestling within a flask; or in the *Donum dei*, where a white queen is succeeded by a red king (Figure 8.14).

Another method of illustrating the sequential nature of chemical operations is to show how matter develops in stages within the flask. This device, already present in Gratheus's poem, became an important theme in fifteenth-century examples, with vessels setting the stage for allegorical scenes in the *Aurora*

FIGURE 8.14 The red elixir, *Donum dei*. Beinecke Rare Books & Manuscript Library, MS Mellon 71 (late seventeenth century), p. 93. By permission of the Beinecke Rare Books & Manuscripts Library.

consurgens, *Donum dei*, *Splendor solis*, and Ripley Scroll. As frames, these alchemical glasses are often rather generic in form; a notable exception being the Ripley Scroll, where specific types of vessel are represented, including a large pelican flask (used for cohobation; see Chapter 2) and several curcurbits with alembics and receivers (for drawing off distillates).

It is the content of the vessel, rather than the apparatus itself, that provides the focus of these sequences, in which the preparation, combination, and transformation of ingredients are depicted as interactions between human and animal figures. In the *Aurora consurgens*, this succession is compactly illustrated by placing three allegorical beasts within a single vessel: below, a dragon, biting its tail to indicate that it is an Ouroboros (and thus signifying the unified prime matter of the work), an eagle (volatilization), and a crow (putrefaction; Gabriele 1997: 77). The same figure appears in the *Holy Trinity* (Figure 8.15).

Other sequences employ a succession of vessels, each stage illustrated by a new rendition of the flask's contents. For instance, the use of an idealized, circular vessel as the framing device for allegorized procedures is the primary conceit of the *Donum dei*, a sequence of twelve images probably dating from the

FIGURE 8.15 Figure from *Das Buch der heiligen Dreifaltigkeit*. Science History Institute, Othmer MS 10b. By permission of the Science History Institute.

late fifteenth century. A series of flasks charts the incremental transformations of the work from prime matter to philosophers' stone, marked by changes in state, color, and symbolism. The main actors in the sequence are the alchemical king and queen who, after coitus, die and putrefy, to be succeeded by worms, a dragon, and eventual triumph as the white and red stones.

THE RIPLEY SCROLL

For all their diversity, the examples of alchemical imagery discussed above still can be considered as accompaniments to (even as commentaries upon) written works, even in cases where they do not obviously illustrate the text. In the Ripley Scroll, one of the most striking productions of fifteenth-century alchemical imagery, this relationship is inverted.[9] Inscribed upon a single parchment roll rather than distributed throughout the pages of a codex, the scale and complexity of the Scroll's design at first seem to overwhelm the English verses tucked around its margins (Figure 8.16). In fact, these alchemical verses now play the role formerly served by illustrative cycles like the *Aurora*, with the text supplying commentary on the visual content.

The earliest surviving witness of the Scroll design is a roll dating to the late fifteenth century (Bodley Rolls 1), although over twenty copies were made during the sixteenth century and later (McCallum 2007; Rampling 2014a). The original design is laid out in a series of panels, conveniently spaced to fit on separate pieces of parchment. The design's use of color symbolism – starting

FIGURE 8.16 The book of seven seals. Ripley Scroll, Princeton University Library MS 93 (late sixteenth century). By permission of the Princeton University Library.

with the black of putrefaction at the head of the roll, then descending through the white and red stages – suggests that the rolls were made to be read from the top down, a conclusion supported by material wear and tear on surviving rolls. The size of the entire design, too large to be conveniently unrolled and viewed at a glance, also suggests that the scroll was intended to be viewed one panel at a time, the process of unrolling and rerolling evoking the seriality of image sequences like the *Donum dei*, but on a grander scale.

The design of the Scroll clearly harks back to an authoritative and ancient past, both through the archaic medium of the roll and the recycling of traditional figures. Yet this is also a work of synthesis: the familiar tropes of Greek and Arabic alchemy, from the chemical wedding to the mercurial dragon, now Christianized and adapted to suit the context of late fifteenth-century England. In a central roundel, a philosopher opens an illustrated book to his disciple, but the emerald tablet described by Ibn Umayl is here replaced by the book of seven seals from the biblical Book of Revelation – a motif that the designer has adapted from an alchemical source, the pseudo-Arnaldian *Visio mystica*, or *Flos florum* (*Secret Vision*, or *Flower of Flowers*). Ibn Umayl's chemical wedding is also reimagined as the temptation of Adam and Eve in the Garden of Eden – the Edenic serpent replaced by a creature that is half-dragon, half-human woman, evoking earlier mercurial figures in *Aurora consurgens* and the *Holy Trinity* (Fgure 8.17).

Alchemical processes are further allegorized using creatures familiar from the alchemical bestiary: the putrefaction of the base matter represented by a dying toad; the dissolution of gold, silver, and mercury by a dragon devouring the sun and moon; and the fixing of the stone by the human-headed "Bird of Hermes" eating its own wings (Rampling 2014a: 41–4).

Through the use of recognizable alchemical figures, even with an additional layer of religious allegory, the Scroll designer signals his commitment to the traditional metallic triumvirate of mercury, gold, and silver, with the addition of a fourth, base metal, represented by a red toad. At the same time, the design reflects a more recent trend in fifteenth-century English alchemy: attention to pseudo-Llullian practices for both transmutional and medicinal ends (Rampling 2020). For instance, the Edenic apple has been replaced by a twisting vine, its grapes a clue to the identity of pseudo-Llull's vegetable stone, made by distilling wine. Still, as with other images, the design of the Scroll can only be fully understood in relation to texts, such as the English verses and the Latin *Visio mystica* – although only the verses are actually incorporated into the Scroll design, leaving the rest of its textual antecedents for learned viewers to discern. These verses, which include variants of the poems later published by Ashmole as the *Work of Pearce the Black Monk* and the *Work of Richard Carpenter*,[10] echo the tropes and doctrines of the pictorial scheme. Only later, as the design was reproduced by early modern copyists (including professional artists who may have lacked specific alchemical expertise), did the accompanying verses fall by the wayside, leaving just the decontextualized images – and a new set of meanings.

FIGURE 8.17 The serpent of Eden. Ripley Scroll, Princeton University Library MS 93. By permission of the Princeton University Library.

CONCLUSION: CYCLES OF ADAPTATION

One of the most striking aspects of the images discussed in this chapter may not be apparent at first glance. Not only do texts often fail to provide any detailed exposition of their accompanying images, but in many cases the images do not obviously relate to the text at all. This discontinuity between text and image is apparent in the *Holy Trinity* and *Aurora consurgens*, but also in later productions like the *Donum dei* and Ripley Scroll. Rather than mapping directly onto one another, text and image describe separate but related narratives. This discontinuity seems to be an example of the technique of *dispersio*, or dispersion of knowledge. According to this technique, already employed in Arabic alchemical writings (*tabdīd al-ʿilm*), necessary information is diffused across diverse texts (Kraus 1943: xxvii–xxx; Newman and Principe 2002: 186–7). Only by reading widely and closely can a reader assemble all of the clues needed to understand the author's meaning.

Such techniques served to promote the authority of alchemy as a body of profound yet secret knowledge that called for special, "philosophical" reading techniques. At the same time, the incorporation of nonalchemical iconographic elements, from the kings, queens, and knights of medieval romance, to religious figures familiar from altarpieces and devotional literature, may have advertised the peculiar epistemological status of alchemy to a wider audience than practitioners alone. The Franciscan author of the *Holy Trinity* dedicated his treatise to the emperor in the hope of drawing attention to his own prophetic message, a manifesto echoed in the interlinking of alchemical, religious, and imperial imagery throughout the work (Obrist 1982: 171–7). More prosaically, the lure of cheap, transmuted bullion and alchemical medicines attracted royal attention in England from the 1450s to the 1470s, when Henry VI of England and his successor Edward IV granted licenses to alchemical practitioners (Rampling 2020: 64–72). This interest left its mark on the matter and form of several alchemical treatises of the period and may have contributed to the reception of manuscripts like the Ripley Scroll – several copies of which include the figure of a king (McCallum 2007: 170–1).

The copying and, eventually, printing of manuscripts would create distance between pictorial elements and the original circumstances of their production, investing them with new forms of authority and permitting new interpretations – shifts that could be magnified by the problems associated with drafting accurate copies. From the intricate trees of pseudo-Llullian alchemy to the figurative components of the Ripley Scroll, alchemical images proved difficult to reproduce, and attempts to do so resulted in considerable variation in quality, detail, and interpretation. In some cases, entire sequences were meticulously copied, including replication of the original color scheme and placement with regard to text. In others, corners were cut and details changed – alterations that might reflect drafting errors, deliberate adaptations, or simply a lack of skill on the part of the copyist. Sometimes attempts at reproduction failed entirely due to miscopying or a failure to leave adequate space on the page – a fate more likely to befall complex diagrams (Rampling 2013: 76–8) (Figure 8.18). Sometimes, scribes left gaps for images, possibly for another artist to complete, that were never filled.

Often, however, alterations to an original design seem to have been deliberate, suggesting copyists' efforts to clarify or interpret their source material, possibly in order to make it legible to an intended audience. Such changes could reflect changes in religious culture, as in the transformation of *The Starry Earth*'s Egyptian pyramid into a Christian church (see Figure 8.10),

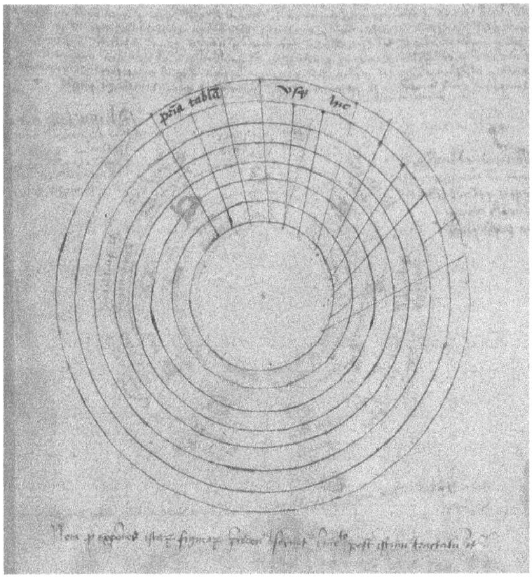

FIGURE 8.18 Failed attempt to plot a pseudo-Llullian figure. Beinecke Rare Books & Manuscript Library, MS Mellon 12, fol. 272v. By permission of the Beinecke Rare Books & Manuscripts Library.

or, in post-Reformation copies, the substitution of figures in secular dress for the Ripley Scroll's original monks. In one version of the Scroll, the grapes in the philosophical tree have even been changed to apples (British Library, Add. MS 5025[3]), a visual graft that brings the design closer to the Eden of Genesis while distancing it from alchemy as actually theorized.

Engagement with earlier sources, melded with the stylistic conventions imported by professional artists, also resulted in entirely new works. The *Splendor solis* (*Splendor of the Sun*), created at the very close of our period in the first half of the sixteenth century, provides one of the most spectacular examples. Whereas the *Aurora consurgens* supplies a partial commentary on a translated Arabic exemplar (the *Tabula chemica*, including some of its figurative elements), the *Splendor solis* engages in turn with the image cycle of the *Aurora*, borrowing selected motifs while also changing salient features to accommodate other source material. For instance, a famous image from the *Aurora* shows the sun and moon (gold and silver) as male and female corpses, beheaded by an ax-wielding, semi-serpentine figure identifiable as mercury (Obrist 1982: 236–7). The *Splendor* includes a similar scene, but omits the figure of silver, while the executioner who brandishes the decapitated head of gold is a bearded character who does not obviously correspond to earlier portrayals of mercury (Figure 8.19). In fact, the accompanying text reveals that the image is intended to illustrate a "parable" attributed to Rosinus (a Latin corruption of Zosimos).

FIGURE 8.19 Decapitation of Sol, *Splendor solis*. Beinecke Rare Books & Manuscript Library, MS Mellon 86, p. 289. By permission of the Beinecke Rare Books & Manuscripts Library.

The deliberate nature of the adaptation is evident even in the work's title, implying a natural, temporal progression from the "Rising Dawn" to the "Sun in Splendor," accompanied by painted illuminations that cast even those of the *Aurora* into the shade (Völlnagel 2004; Völlnagel 2011) (Figure 8.20). In the image cycle, the comparison is underscored by a dramatic transition from the penultimate image – the sun rising from behind a shadowy line of hills – to a closing vision of the sun in majesty above a fully illuminated landscape.

FIGURE 8.20 The risen sun, *Splendor solis*. British Library, Harley MS 3469 (1582), fol. 33v.

Even as it glances back to fifteenth-century models, the *Splendor* is already leaning in to new artistic currents. Dress, pose, architecture, and landscape all reflect the classical ideals of Renaissance art – a wrangling of medieval tropes into humanist dress that would come to characterize much of the alchemical art and literature produced in the sixteenth century and beyond. Such stylistic renovations point, yet again, to the durability and flexibility of alchemical visualizations, which, like text, furnished endless scope for remodeling and reinterpretation. In designing new cycles, image-makers offered both an echo and an elaboration of earlier traditions, adopting, modifying, and combining familiar elements to offer a visual commentary not just on textual authorities, but also on previous attempts to visualize alchemical knowledge. The matter of these figures did not remain fixed but, like mercury itself, transmuted into new forms and new styles – a fitting enough allegory for the art of fire.

NOTES

INTRODUCTION

1 Research for this chapter benefited from the support of the FNRS and the ERC project "The Origin and Early Development of Philosophy in Tenth-century al-Andalus: the Impact of Ill-defined Materials and Channels of Transmission" (ERC 2016, AdG 740618, PI Godefroid de Callataÿ) at the University of Louvain (Université catholique de Louvain), from 2017 to 2022.

2 Marion Dapsens is currently preparing a critical edition of the Latin and Arabic texts.

3 The critical edition of the Arabic and Hebrew versions is currently being prepared by Gabriele Ferrario and the critical edition of the Latin version is currently being prepared by Catherine Arbuthnott.

4 This phenomenon leads the reader to a second degree of pseudepigraphy: these are texts written in Latin attributed to an Arabic author's name, Jābir b. Ḥayyān, so they are pseudepigraphs, but the Arabic texts that carry the name Jābir are themselves in all likelihood pseudepigraphs, since Jābir may never have existed and certainly did not write everything attributed to him.

CHAPTER 1

1 Research for this chapter benefited from the support of the FNRS, the ERC project "The Origin and Early Development of Philosophy in Tenth-century al-Andalus: The Impact of Ill-defined Materials and Channels of Transmission" (ERC 2016, AdG 740618, PI Godefroid de Callataÿ) at the University of Louvain (Université catholique de Louvain), from 2017 to 2022, and the ERC project "Alchemy in the Making: From ancient Babylonia via Graeco-Roman Egypt into the Byzantine, Syriac and Arabic traditions (1500 BCE – 1000 AD)" (CoG 724914, PI Matteo Martelli) at the Università di Bologna, from 2017–2022.

2 See, for instance, books I and II of Aëtius' *Libri medicinales* (devoted to simple medicines) and the seventh book of Paul's *Medical Compendium* (Gowling 2017: 85–9, on Aëtius; Scarborough 1984: 228–9, on Paul).

3 See Viano (2006: 211–23) on Olympiodorus; see chapter 23, sections 7–9 of Blemmydes' *Epitome physica* (*PG* 142: 1211–14).

4 See also Stephanus, *Letter to Theodoros*, ll. 13–20 (Papathanassiou 2017: 167 = Taylor 1938: 38–9) and the philosopher Christianus (Berthelot and Ruelle 1887–1888: II:415–16).
5 See, for instance, Papathanassiou (2005: 123–7) on Stephanus of Alexandria and Stephanus, *Lecture* IV 81–125 (Papathanassiou 2017: 175–7); *Lecture* VIII 103–16 (Papathanassiou 2017: 210–1).
6 Already in the *Book of the Secret of Creation* (*Kitāb sirr al-khalīqa*) by Pseudo-Apollonius of Tyana (ninth century): see, for instance, Weisser (1979: 231–2). For this theory in the Jābirian corpus, see Kraus (1942: 1–2). This is discussed further below, under the classification of substances.
7 Amount and not intensity, since the properties are considered as kinds of elements prior to elements, contrary to what one reads in Aristotle. In the Jābirian *Kitāb al-aḥjār 'alā ra'y Balīnās*, the author speaks of *qīrāṭ* (of properties), *qīrāṭ* being a measure of weight.
8 For a detailed description of the system, see Kraus (1942: 187–303).
9 On "mercury alone," see Thorndike (1923–58, III:58, 89–90); Principe (1998: 153–5). William Newman argues that the approach originated in the *Summa perfectionis* (1991: 204–8). On the material status of "mercury," see Newman (2014).
10 Pseudo-Democritus instead classified natural substances according to their use and consistency: see Martelli (2013: 26–9).
11 Olympiodorus, *In Aristotelis Meteora*, III 6 (Stüve 1900: 266–7). See Halleux (1973: 149–60).
12 Already in the Stoic tradition, philosophy was likened to an egg: ethics to its yolk, physics to the white, logic to the shell (Diogenes Laertius, *Lives and Opinions of Eminent Philosophers*, VII 40).
13 Similar associations (with no mention of the four elements) can be found in another Byzantine text on the names of eggs: see Berthelot and Ruelle (1887–1888: II:18–20).
14 These classes were further divided into subgroups by the philosopher Christianus, whose classification was guided by a complex numerology and ended up establishing 135 alchemical combinations.
15 Psellus, *Letter on the Making of Gold*, §§ 1–5 (Bidez 1928: 26–33).
16 See, for instance, Berthelot and Ruelle (1887–1888: II:399–408) on the philosopher Christianus, and Berthelot and Ruelle (1887–1888: II:421–33) on the *Anepigraphos* philosopher.
17 Martelli (2013: 132–5 [Syn. §§ 8–9] and 245–7); Dufault (2015: 230–8).
18 Bacon, *Opus tertium* (Brewer 1859: 39–40). The foundational role of alchemy in Bacon's natural philosophy has often been noted: see, inter alia, Pereira (1992: 54–5); Newman (1994a: 461–2).
19 Unfortunately, there is no clear evidence of transmission between Bacon's theories and these fourteenth-century works: see, however, Pereira (1992, 1995, 1998) and Crisciani (forthcoming).
20 On the pseudo-Arnaldian corpus, see Calvet (2011); on pseudo-Llull, see Pereira (1989, 1992); on Dastin, see Calvet (2011: 128–30) and Rodríguez Guerrero (2010–13: 92–101).

CHAPTER 2

1 Research for this chapter benefited from the support of the FNRS and the ERC project "The Origin and Early Development of Philosophy in Tenth-century al-Andalus: the Impact of Ill-defined Materials and Channels of Transmission"

(ERC 2016, AdG 740618, PI Godefroid de Callataÿ) at the University of Louvain (Université catholique de Louvain), from 2017 to 2022, and of the IN WBI (Wallonie – Bruxelles International) program.

2 The structure of Rāzī's work is explained in two places: the seven chapters are described at the beginning of the book (Dānish-Pazhūh 1964: 13), while the titles of the subsections are found across the entire work (pp. 13–116).

3 Actually, the first part of the first section is dedicated to operations on spirits (i.e. mercury, sulfur, ammoniac salt, and arsenic – actually arsenic sulfides; see Chapter 1). For reasons of space, we cannot address the question of the preparation of spirits here, but the coagulation of mercury and the sublimation of spirits will be explained below.

4 *Rectificatio* might, however, have another meaning, especially in texts of the early period of Latin alchemy.

5 Rāzī mentions it but does not explain it in detail.

6 For instance, the author of the *Summa perfectionis* mentions the two operations in two different chapters, but explains that the distillation *per descensum* is made in the *chimina* (i.e. the descensory), acknowledging a similarity in the devices (*cap.* 46; Newman 1991: 408). In modern classification, descension is related to mineral chemistry, while distillation *per descensum* is related to organic chemistry.

7 Another solution, when gold is impure, is to mix it with sulfur compounds, which will react with the other metals mixed with the gold.

8 This result of the sublimation is actually in most cases a mixture of corrosive sublimate ($HgCl_2$) and calomel (Hg_2Cl_2).

9 The merchant Pegolotti talks about *lasagna* to explain the system of superimposing layers of cement and metal (see Evans 1936: 331–3).

CHAPTER 3

1 Research for this chapter benefited from the support of the FNRS and the ERC project "The Origin and Early Development of Philosophy in Tenth-century al-Andalus: the Impact of Ill-defined Materials and Channels of Transmission" (ERC 2016, AdG 740618, PI Godefroid de Callataÿ) at the University of Louvain (Université catholique de Louvain), from 2017 to 2022, and of the IN WBI (Wallonie – Bruxelles International) program.

2 Crafts workshops and tools will be addressed in Chapter 6.

3 We thank Antoine Calvet for his analysis of this manuscript.

4 For distillation *per descensum*, see examples in the French translation of the *Liber de simplici medicina dictus Circa instans* of Platearius (Dorveaux 1913) or English recipes in Moorhouse (1981).

5 As a result of the scarcity of Arabic alchemical iconography, we had to use Latin texts to better understand the Arabic aludel (Moureau and Thomas 2015), and Sezgin and Neubauer, when trying to replicate alchemical tools and furnaces, had to use Latin materials (see Sezgin and Neubauer 2010 to read *cum grano salis* since some replicas have practical inconsistencies).

6 See, for example, the *Liber de arte distillandi de simplicibus* of Hieronymus Brunschwig, written in German and published in Strasbourg in 1500.

7 Archaeological remains dated after 1500 are not considered in this book. These remains are very numerous in Europe, such as the remains in Oberstockstall, Kirchberg am Wagram (Austria), which have been the subject of many studies, some of them general, others more specific (Soukup and Mayer 1997; Osten 1998); for a recent bibliography, see Meller et al. (2016).

CHAPTER 4

1 The section *Islamicate World* was written in the framework of the project "Between Religion and Alchemy: The Scholar Ibn Arfaʿ Raʾs (d. 1197) as a Model for an Integrative Arabic Literary and Cultural History," funded by the Swiss National Science Foundation.

2 The differentiation into *ʿulūm ʿaqliyya* ("rational sciences") and *ʿulūm naqliyya* ("sciences relying on traditions") seems not to be alluded to when it comes to categorizing alchemy; on this differentiation, see Rudolph (2012a: xxx–xxxi).

3 I should like to thank Sébastien Moureau for referring me to this passage.

4 The prominence of mercury in Islamicate alchemy might be due to Indian influence. On the use of mercury in Indian alchemy, see Hellwig (2009: esp. 12–13, 306–13).

5 Al-Jildakī probably was an Egyptian-born descendant of Turkic Mamlūks (Harris 2017: 556); therefore, this form of his name seems to be preferable to the Persian form al-Jaldakī adapted by Corbin (1986: 67). Dates of his life are uncertain, but he was alive after 743 AH/1342 CE, the date usually given as year of his death (see Forster and Müller 2020).

6 It is interesting to note that Ibn al-Ḥājj seems to have offered a different view on alchemy in a treatise entitled *al-Nukta* ("The witty point"), extant in a manuscript at University of California, Los Angeles (Ar. 17, fols 65–77); see Iskandar (1984: esp. 23).

7 The monograph by Raphael Patai on Jewish alchemists (1994) draws a very inclusive (but not equally solid) portrait of the intersections of Judaism and alchemy. Reviews of this work (Freudenthal 1995; Langermann 1996) have highlighted its shortcomings.

8 The figure of Mary occupies two chapters of Patai's monograph (1994: 60–91), where a large number of quotations from works by Zosimos are translated and commented upon. On the original works by Zosimos, see Mertens (1995).

9 Freudenthal's 2011 article is currently the most complete and balanced survey of all the available information on the involvement of Jews in alchemy during the Middle Ages. Freudenthal points out a number of areas of research (and of alchemical manuscripts and corpora) that may yield further important information on the topic but have not been sufficiently studied yet. Work is currently underway for the publication of the alchemical corpus of the Cairo Genizah (see below).

10 I am here providing an abridged overview of the sources that are already discussed in Freudenthal (2011), and the reader should refer to that work for further details. I am integrating Freudenthal's information with evidence derived from my own work on the Cairo Genizah collections.

11 In this document, the fraudulent alchemist is also called a "money exchanger" and the same identification is found in two other eleventh-century court records from the Cairo Genizah: Cambridge University Library T-S 18J1.11 and T-S 12.1. On the first fragment, see Golb et al. (1958).

12 I have recently completed a survey of the alchemical fragments preserved in the Genizah Collection at Cambridge University Library that has revealed the presence of more than 100 manuscripts of alchemical content, totaling more than 350 pages of text. I intend to edit and publish a selection of these texts.

13 The fragments are, in order: Cambridge University Library, T-S Ar.35.104, T-S Ar.53.58, T-S Ar.43.267, T-S AS 160.251 (Bink Hallum firstly identified these fragments and kindly shared his discovery with me), T-S NS 31.6, T-S NS 31.6e, and T-S Ar.44.4.

14 This is the hypothesis that I have advanced in a number of scholarly meetings and workshops; it still needs to be backed by further and stronger evidence.
15 The alchemical treatise occupies the margins of eighteen folios of manuscript Toledo, Archivo y Biblioteca Capitulares, MS Z-98-13.
16 As noted by Freudenthal (2011: 354–5), these interesting figures of alchemists and their relationships with rulers and patrons, local authorities, and their Jewish communities of belonging deserve more scholarly attention.
17 The most important of these manuscripts are: London, British Library, Or. 10289 (known as the Gaster manuscript); Manchester, John Rylands University Library, No. 1435; and Berlin, Staatsbibliothek, Or. Oct. 514. A description of their content, together with the translation of few excerpts, is given in Patai (1994: 420–34 [London manuscript]; 381–92 [Manchester manuscript]; 407–16 [Berlin manuscript]).
18 For reasons of space and expertise, I will limit this survey to a summary of the evidence highlighted in Freudenthal (2011), on which this section is based.
19 Freudenthal (2011: 349) argues that Gershon's careful avoidance of embracing the pro-transmutational positions of the *Emm Ha-Melekh* might have been caused by his awareness, through Christian intermediaries, of the Latin version of Aristotle's *Meteorologica IV*, which included an addition from Ibn Sīnā's *Shifā'* that denied the possibility of transmutation.

CHAPTER 5

1 This is the genuine Llull rather than the Pseudo-Llull to whom several alchemical texts were attributed.
2 For the association of Matthias Corvinus (king of Hungary,1458–90) with alchemy, see Scafi (1993).
3 "*quoniam tantum de vero auro et argento debent inferre in publicum ut pauperibus erogetur, quantum de falso et adulterino posuerunt*." Also John XXII's *Extravagantes* XX (Lyons, 1535): *De crimine falsi*. Ioan. xxii.

CHAPTER 6

1 Both Theophilus (Hawthorne and Smith 1979: 109) and Agricola (Hoover and Hoover 1950: 439) provide good descriptions of the process.
2 Theophilus does not mention it, but Ercker (Sisco and Smith 1951) and Agricola do. For the process and illustrations of the vessels used, see Chapter 2.
3 Most notably in the *Ta'rīkh al-rusul wa'l-mulūk* ("History of Prophets and Kings") of al-Ṭabarī (838–923 CE).
4 Theophilus' *De diversis artibus*, Ch. 66, contains a good description of this process.

CHAPTER 7

1 The section *Islamicate World* was written in the framework of the project "Between Religion and Alchemy: The Scholar Ibn Arfa' Ra's (d. 1197) as a Model for an Integrative Arabic Literary and Cultural History," funded by the Swiss National Science Foundation.

2 This commentary survives in a rather large number of manuscripts representing different recensions of the text, but there is, against Ullmann's assumption (1972: 232), only one commentary by Ibn Arfaʿ Raʾs himself. Juliane Müller (Zurich) will publish an edition of the commentary in Beirut (Bibliotheca Islamica) soon.

3 I wish to thank Svetlana Dolgusheva (Zurich), who is currently preparing a critical edition of *Shudhūr al-dahab*, for providing me with a list of such passages.

4 https://www.qdl.qa/archive/81055/vdc_100000003591.0x000001 (Accessed February 5, 2018). I should like to thank Bink Hallum (London) for sharing this reference with me.

5 For a general survey of the history of medieval alchemical literature, see Calvet (2018a), and of medieval alchemical poetry, see Kahn (2010a) and Kahn (2011).

6 See, for instance, the anonymous fourteenth-century genealogy of alchemy bearing the title "Who Were the First Discoverers of This Art" (*Qui fuerint primi inventores hujus artis*, ed. Kahn 2017); see other examples in Mandosio (2000).

CHAPTER 8

1 I am grateful to Bink Hallum and Sébastien Moureau for their invaluable suggestions and advice regarding the Arabic sources discussed in this chapter.

2 For the purposes of this chapter, I use "image" as a generic term referring to modes of visual representation. On the problems associated with the classification of medieval diagrams (including terms like "figure" and "diagram"), see North (2004: 265–87) and Lüthy and Smets (2009: 420–4).

3 Since drawings of this type are discussed in detail in Chapter 3, I do not examine them further here.

4 See, for instance, MS Paris, BnF, arabe 5099, on fols 73v, 76r, and 97r–107r.

5 See, for instance, the diagrams in volumes I and IV of Bethesda, MD, National Medical Library, shelfmark A7. I am most grateful to Nicholas Harris for suggesting this example.

6 İstanbul Arkeoloji Muzeleri Kütüphanesi, MS 1574. The images evidently accompanied the text in earlier versions, since Maslama discusses them in the *Rutba* (tenth century CE) and claims to have written an entire book on the subject (Callataÿ and Moureau forthcoming).

7 British Library, Add. MS 25724, f. 18r. Although permission to reproduce these images was not granted for this volume, several digital reproductions can be consulted in Hallum and Marée 2016.

8 Telle (1980: 45–54). In 1550, the sequence was translated into a famous series of woodcut images to accompany the Latin florilegium, the *Rosarium philosophorum* (*The Rosegarden of Philosophers*, or *Rosary of Philosophers*, 1550; Telle 1992).

9 Although George Ripley's name became attached to the Scroll and its verses during the sixteenth century, this seems to have been a later attribution (McCallum 2007; Rampling 2010; Timmermann 2013).

10 Ashmole (1652: 269–77). They have been more recently edited by Timmermann (2013: 215–32, 266–85), together with a partial edition of the Scroll verses (pp. 294–303).

BIBLIOGRAPHY

'Abd al-Ḥamīd, Muḥammad Muḥyī l-Dīn (ed.). 1951–1953. Muḥammad b. Shākir al-Kutubī, *Fawāt al-wafayāt*. 2 vols. Cairo: Maktabat al-Nahḍa al-Miṣriyya.

Abrahams, Harold J. 1984. "Al-Jawbari on False Alchemists." *Ambix*, 31: 84–8.

Abt, Theodor. 2007a. *The Book of Pictures. Muṣḥaf aṣ-ṣuwar by Zosimos of Panopolis. Edition of the Pictures and Introduction* (Corpus Alchemicum Arabicum, Supplement). Zürich: Living Human Heritage Publications.

Abt, Theodor. 2007b. *The Book of Pictures. Muṣḥaf aṣ-ṣuwar by Zosimos of Panopolis. Facsimile with an Introduction* (Corpus Alchemicum Arabicum, II.1). Zürich: Living Human Heritage Publications.

Abt, Theodor, and Salwa Fuad. 2011. *The Book of Pictures. Muṣḥaf aṣ-ṣuwar by Zosimos of Panopolis. Introduction and Translation* (Corpus Alchemicum Arabicum, II.2). Zürich: Living Human Heritage Publications.

Addas, Claude. 1993. *Quest for the Red Sulphur. The Life of Ibn 'Arabī*. Transl. from the French by Peter Kingsley. Cambridge: The Islamic Texts Society.

Adlington, Laura Ware, Ian Charles Freestone, Jerzy J. Kunicki-Goldfinger, Tim Ayers, Heather Gilderdale Scott, and Anna Eavis. 2019. "Regional Patterns in Medieval European Glass Composition as a Provenancing Tool." *Journal of Archaeological Science*, 110: 104991.

Africanus, Leo. 1830. *De l'Afrique, contenant la description de ce pays par Léon l'Africain, et la navigation des anciens capitaines portugais aux Indes orientales et occidentales*. Paris: De L. Cordier/De Ducessois.

Aït Salah Semlali, Kacem. 2015. *Histoire de l'alchimie et des alchimistes au Maroc*. S.l.: published privately.

Alexander of Hales. 1960. *Quaestiones disputatae "antequam esset frater"* (Bibliotheca Franciscana scholastica Medii Aevi, 19–21). Rome: Collegio S. Bonaventura.

Allemann, Franz. 1988. "'Abdallaṭīf al-Baġdādī. Ris. fī Muġādalat al-ḥakīmain al-kīmiyā'ī wan-naẓarī ('Das Streitgespräch zwischen dem Alchemisten und dem theoretischen Philosophen'). Eine textkritische Bearbeitung des Handschrift: Bursa, Hüseyin Çelebi 823, fol. 100–123 mit Übersetzung und Kommentar." Ph.D. thesis, University of Bern.

Amadori, Gabriele (ed.). 2014. Leo Africanus, *Cosmographia de l'Affrica (Ms. V.E. 953 – Biblioteca Nazionale Centrale di Roma – 1526)*. Rome: Aracne.

Anawati, Georges C. 1996. "Arabic Alchemy." In Roshdi Rashed and Régis Morelon (eds), *Encyclopedia of the History of Arabic Science. Vol. 3. Technology, Alchemy and Life Sciences*. London: Routledge.

Anderson, Robert G.W. 2000. "The Archaeology of Chemistry." In Frederic L. Holmes and Trevor H. Levere (eds), *Instruments and Experimentation in the History of Chemistry*. Cambridge, MA: MIT Press.

Anonymous. 1886. "Continuatio Thomae de Aquino Expositionis in Aristotelis libros Meteorologicorum." In *Opera omnia: iussu impensaque Leonis XIII. P.M. edita (= editio Leonina). 3 Commentaria in libros Aristotelis De caelo et mundo, De generatione et corruptione, et Meteorologicorum*. Rome: Typographia Polyglotta S.C. de Propaganda Fide.

Arbuthnott, Catherine (ed.). Forthcoming. Pseudo-Rāzī. *De aluminibus et salibus*.

Artun, Tuna. 2013. "Hearts of Gold and Silver: The Production of Alchemical Knowledge in the Early Modern Ottoman World." Ph.D. thesis, Princeton University.

Ashmole, Elias (ed.). 1652. *Theatrum chemicum Britannicum*. London: Nathaniel Brooke.

Assaf, Simcha. 1942. *Responsa Geonica*. Jerusalem: Mekize Nirdamim [in Hebrew].

Awty, Brian. 2003. "The Queenstock Furnace at Buxted, Sussex: The Earliest in England?" *Historical Metallurgy*, 37: 51–2.

Baalbaki, Ramzi (ed.). 1983. *Al-Ṣafadī, Ṣalāh al-Dīn Khalīl b. Aybak, Kitāb al-Wāfī bi-l-wafayāt. Vol. 22: 'Alī b. Muḥammad b. Rustam to 'Umar b. 'Abd an-Naṣīr.* Wiesbaden: Steiner.

Bacchi, Eleonora, and Matteo Martelli. 2009. "Il principe Ḫālid bin Yazīd e le origini dell'alchimia araba." In Daniele Cevenini and Svevo D'Onofrio (eds), *'Uyūn al-Akhbār. 3. Conflitti e dissensi nell'Islam*. Bologna: Il Ponte.

Bachoffner, Pierre. 1993. "Jérôme Brunschwig, chirurgien et apothicaire strasbourgeois, portraituré en 1512." *Revue d'histore de la pharmacie*, 298: 269–78.

Bachour, Natalia. 2012. *Oswaldus Crollius und Daniel Sennert im frühneuzeitlichen Istanbul. Studien zur Rezeption des Paracelsismus im Werk des osmanischen Arztes Ṣāliḥ b. Naṣrullāh Ibn Sallūm al-Ḥalabī*. Freiburg: Centaurus.

Barthélémy, Pascale. 1995. "Le verre dans la *Sedacina totius artis alchimie* de Guillaume Sedacer." In Didier Kahn and Sylvain Matton (eds), *Alchimie: art, histoire et mythes. Actes du Ier colloque international de la Société d'Étude de l'Histoire de l'Alchimie, Paris, Collège de France, 14–16 mars 1991*. Paris, Milan: S.É.H.A.-Archè.

Barthélémy, Pascale (ed.). 2002. *La Sedacina ou l'Œuvre au crible : l'alchimie de Guillaume Sedacer, carme catalan de la fin du XIVe siècle* (Textes et Travaux de Chrysopœia, 8). Paris: S.É.H.A.-Archè.

Bauemker, Clemens (ed.). 1916. Alfarabi, *Über der Ursprung der Wissenschaften (De ortu scientiarum): Eine mittelalterliche Einleitungsschrift in die philosophischen Wissenschaften*. Münster: Aschendorff.

Baumgartner, Erwin, and Ingeborg Krueger. 1988. *Phönix aus Sand und Asche*. Munich: Klinkhardt & Biermann.

Bayley, Justine. 1992. *Non-ferrous Metalworking at 16–22 Coppergate* (The Archaeology of York, 17/7). London: Council for British Archaeology.

Bayley, Justine. 1998. "The Production of Brass in Antiquity with Particular Reference to Roman Britain." In Paul T. Craddock (ed.), *2000 Years of Zinc and Brass* (Occasional Papers, 50). Rev. edn. London: British Museum.

Bayley, Justine. 2008, "Medieval Precious Metal Refining: Archaeology and Contemporary Texts Compared." In Marcos Martinón-Torres and Thilo Rehren (eds), *Archaeology, History and Science: Integrating Approaches to Ancient Materials* (University College London Institute of Archaeology Publications). Walnut Creek, CA: Left Coast Press.

Bayley, Justine. 2009. "Early Medieval Lead-Rich Glass in the British Isles – A Survey of the Evidence." In Koen Janssens, Patrick Degryse, Peter Cosyns, Joost Caen, and Luc Van't Dack (eds), *Annales du 17ᵉ Congrès de l'Association Internationale pour l'Histoire du Verre*. Antwerp: University Press Antwerp.

Bayley, Justine, and Andy Russel. 2008. "Making Gold–Mercury Amalgam: The Evidence for Gilding from Southampton." *Antiquaries Journal*, 88: 37–42.

Bayley, Justine, and Harriet White. 2013. "Evidence for Workshop Practices at the Tudor Mint in the Tower Of London." In David Saunders, Marika Spring, and Andrew Meek (eds), *The Renaissance Workshop*. London: Archetype Publications.

Bayley, Justine, and Kerstin Eckstein. 1997. "Silver Refining – Production, Recycling, Assaying." In Anthony Sinclair, Elizabeth Slater, and John Gowlett (eds), *Archaeological Sciences 1995*. Oxford: Oxbow Books.

Bayley, Justine, and Kerstin Eckstein. 2006. "Roman and Medieval Litharge Cakes: Structure and Composition." In *Proceedings of the 34th International Symposium on Archaeometry, Zaragoza, 2004*. Zaragoza: Institución Fernando El Católico.

Beaujouan, Guy, and Paul Cattin. 1981. "Philippe Éléphant (mathématique, alchimie, éthique)." In *Histoire littéraire de la France. Tome XLI, suite du quatorzième siècle*. Paris: Imprimerie nationale.

Bénézet, Jean-Pierre. 1999. *Pharmacie et médicament en Méditerranée occidentale (XIIIᵉ–XVIᵉ siècles)* (Sciences, techniques et civilisations du Moyen Age à l'aube des Lumières, 3). Paris: Honoré Champion.

Beretta, Marco. 2002. *Storia materiale della scienza: dal libro ai laboratori*. Milan: Mondadori.

Berlekamp, Persis. 2003. "Painting as Persuasion: A Visual Defense of Alchemy in an Islamic Manuscript of the Mongol Period." *Muqarnas*, 20: 35–59.

Berthelot, Marcellin. 1906. "Archéologie et histoire des sciences." *Mémoires de l'Académie des sciences de l'Institut de France*, 49(2): 1–377.

Berthelot, Marcellin, and Charles-Émile Ruelle. 1887–1888. *Collection des anciens alchimistes grecs*. Paris: G. Steinheil.

Berthelot, Marcellin, Octave Victor Houdas, and Rubens Duval. 1893. *La chimie au Moyen Âge*. Paris: Imprimerie nationale.

Besborodov, Mikhail. A. 1957. "A Chemical and Technological Study of Ancient Russian Glasses and Refractories." *Journal of the Society of Glass Technology*, 41: 168–84.

Bidez, Joseph. 1928. Michel Psellus. *Épître sur la Chrysopée, opuscules et extraits sur l'alchimie, la météorologie et la démonologie* (Catalogue des manuscrits alchimiques grecs, vol. 6). Brussels: M. Lamertin.

Biek, Leo, and Justine Bayley. 1979. "Glass and Other Vitreous Materials." *World Archaeology*, 10: 1–25.

Birkhan, Helmut. 1992. *Die alchemistische Lehrdichtung des Gratheus filius philosophi in cod. Vind. 2372. Ein Beitrag zur okkulten Wissenschaft im Spätmittelalter*. 2 vols. Vienna: Österreichische Akademie der Wissenschaften.

Boemer, Alois (ed.). 1924. *Epistulae obscurorum virorum*. Heidelberg: R. Weissbach.

Borgnet, Auguste (ed.). 1890. Albert le Grand. *Beati Alberti Magni, Ratisbonensis Episcopi, ordinis praedictorum. Opera omnia. Vol.* 5. Paris: Louis Vivès.

Borgnet, Auguste, and Emile Borgnet (eds). 1908. Albert le Grand. *Beati Alberti Magni, Ratisbonensis Episcopi, ordinis praedictorum. Opera omnia. Vol.* 37. Paris: Louis Vivès.

Bos, Gerrit. 1994. "Hayyim Vital's Kabbalah Ma'asit we-Alkhimiyah (Practical Kabbalah and Alchemy), a Seventeenth-Century 'Book of Secrets.'" *Journal of Jewish Thought and Philosophy*, 4: 55–112.

Bougerol, Jacques Guy (ed.). 1993. Bonaventura. *Sermones de diversis*. 2 vols. Paris: Editions Franciscaines.

Bourlet, Caroline, and Nicolas Thomas. 2018. "Les métiers du cuivre à Paris vers 1300: topographie et étude sociale." In Nicolas Thomas and Pete Dandridge (eds), *Cuivre, bronzes et laitons médiévaux/Medieval copper, bronze and brass* (Études et documents, 39). Namur: Agence wallonne du Patrimoine.

Braun, Christopher. 2016. *Das Kitāb Sidrat al-muntahā des Pseudo-Ibn Waḥshiyya: Einleitung, Edition und Übersetzung eines hermetisch-allegorischen Traktats zur Alchemie* (Islamkundliche Untersuchungen, 327). Berlin: Klaus Schwarz Verlag.

Braun, Christopher. 2017. "Treasure Hunting and Grave Robbery in Islamic Egypt. Textual Evidence and Social Context." Ph.D. thesis, Warburg Institute, School of Advanced Study, University of London.

Brepohl, Erhard. 1999. *Theophilus Presbyter und das mittelalterliche Kunsthandwerk. Gesamtausgabe der Schrift De diversis artibus in zwei Bänden*. Cologne: Böhlau.

Brewer, John S. (ed.). 1859. Roger Bacon, *Opus tertium*. In *Fratris Rogeri Bacon opera quaedam hactenus inedita*, vol. 1. London: Longman.

Bridges, John Henry (ed.). 1897. Roger Bacon, *Opus maius*. 2 vols. Oxford: Clarendon Press.

Brill, Robert H. 2005. "Chemical Analyses of Some Sasanian Glasses from Iraq." In David Whitehouse (ed.), *Sasanian and Post-Sasanian Glass in the Corning Museum of Glass*. Corning, NY: Corning Museum of Glass.

Brill, Robert H. 2009. "Chemical analyses." In George F. Bass, Robert H. Brill, Berta Lledó, and Sheila D. Matthews (eds), *Serçe Limanı, Vol. II: The Glass of an Eleventh-Century Shipwreck*. College Station: Texas A&M University Press.

Brockelmann, Carl. 1937–1949. *Geschichte der arabischen Litteratur*, 2nd ed. 3 vols and 2 supplement vols. Leiden: Brill.

Brown, James Wood. 1897. *An Enquiry into the Life and Legend of Michael Scot*. Edinburgh: David Douglas.

Brunschwig, Hieronymus. 1500. *Liber de arte distillandi, de simplicibus: Das Buch der rechten Kunst zu distilieren die eintzige[n] Ding*. Strasbourg: Johan Grüninger.

Buchanan, Peter T. 1992. "Metalliferous Plan Communities: The Flora of Lead Smelting in the Upper Nent Valley." In Lynn Willies and David Cranstone (eds), *Boles and Smelt Mills*. London: Historical Metallurgy Society.

Burnett, Charles. 1992. "The Astrologer's Assay of the Alchemist: Early References to Alchemy in Arabic and Latin Texts." *Ambix*, 39: 103–9.

Burnett, Charles. 2001. "The Coherence of the Arabic–Latin Translation Program in Toledo in the Twelfth Century." *Science in Context*, 14: 249–88.

Burnett, Charles. 2002. "The Strategy of Revision in the Arabic–Latin Translations from Toledo: The Case of Abū Ma'shar's *On the Great Conjunctions*." In Jacqueline Hamesse (ed.), *Les traducteurs au travail: leurs manuscrits et leurs méthodes*. Turnhout: Brepols.

Burnett, Charles. 2008. "Translation from Arabic into Latin in the Middle Ages." In Harald Kittel et al. (eds), *Übersetzung. Translation. Traduction. Encyclopédie internationale de la recherche sur la traduction*. Berlin: Walter de Gruyter.

Burnett, Charles. 2009. *Arabic into Latin in the Middle Ages. The Translators and Their Intellectual and Social Context*. Farnham: Ashgate.

Burnett, Charles. 2013. "Translation and Transmission of Greek and Islamic Science to Latin Christendom." In David C. Lindberg and Michael H. Shank (eds), *The Cambridge History of Science. Vol. 2. Medieval Science*. Cambridge: Cambridge University Press.

al-Bustānī, Buṭrus (ed.). 1957. *Rasā'il Ikhwān al-Ṣafā'*. Beirut: Dār Ṣādir.

Buttimer, Charles Henry (ed.). 1939. Hugh of St Victor, *Didascalicon de studio legendi*. Washington, DC: Catholic University Press.

Callataÿ, Godefroid de. 2013. "Magia en al-Andalus: *Rasā'il Ijwān al-Ṣafā', Rutbat al-Ḥakīm* y *Gāyat al-Ḥakīm* (*Picatrix*)." *Al-Qanṭara*, 34: 297–343.

Callataÿ, Godefroid de, and Bruno Halflants. 2011. *On Magic, 1. An Arabic Critical Edition and English Translation of Epistle 52a*. Oxford: Oxford University Press in association with The Institute of Ismaili Studies.

Callataÿ, Godefroid de, and Sébastien Moureau. 2015. "Towards the Critical Edition of the *Rutbat al-ḥakīm*: A Few Preliminary Observations." *Arabica*, 62: 385–94.

Callataÿ, Godefroid de, and Sébastien Moureau. 2016. "Again on Maslama Ibn Qāsim al-Qurṭubī, the Ikhwān al-Ṣafā' ... and Ibn Khaldūn: New Evidence from Two Manuscripts of the *Rutbat al-ḥakīm*." *Al-Qanṭara*, 37: 329–72.

Callataÿ, Godefroid de, and Sébastien Moureau. 2017. "A Milestone in the History of Andalusī Bāṭinism: Maslama b. Qāsim al-Qurṭubī's *Riḥla* in the East." *Intellectual History of the Islamicate World*, 5: 86–117.

Callataÿ, Godefroid de, and Sébastien Moureau. Forthcoming. "In Code We Trust. The Concept of *Rumūz* in Andalusī Alchemical Literature and Related Texts." *Asiatische Studien*, 75.

Calvet, Antoine. 2006. "Étude d'un texte alchimique latin du XIVe siècle: le *Rosarius philosophorum* attribué au médecin Arnaud de Villeneuve (*ob*. 1311)." *Early Science and Medicine*, 11: 162–206.

Calvet, Antoine. 2011. *Les Œuvres alchimiques attribuées à Arnaud de Villeneuve. Grand œuvre, médecine et prophétie au Moyen Âge* (Textes et Travaux de Chrysopœia, 11). Paris: S.É.H.A.-ARCHÈ.

Calvet, Antoine. 2012. "L'alchimie du Pseudo-Albert le Grand." *Archives d'Histoire Doctrinale et Littéraire du Moyen Âge*, 79: 115–60.

Calvet, Antoine. 2017. "L'influence du *Roman de la Rose* dans les textes alchimiques des XIVe et XVe siècles." In Jean-Patrice Boudet, Philippe Haugeard, Silvère Menegaldo, and François Ploton-Nicollet (eds), *Jean de Meun et la culture médiévale: littérature, art, sciences et droit aux derniers siècles du Moyen* Âge. Rennes: Presses universitaires de Rennes.

Calvet, Antoine. 2018a. *L'alchimie au Moyen Âge (XIIe–XVe siècles)* (Études de philosophie médiévale, 107). Paris: Vrin.

Calvet, Antoine. 2018b. "Le *De arte alchemica* (*inc.* Dixit Arturus explicator huius operis') est-il une œuvre authentique de Richard de Fournival?" In Joëlle Ducos and Christopher Lucken (eds), *Richard de Fournival et les sciences au XIIIe siècle* (Micrologus' Library, 88). Florence: Edizioni del Galluzzo.

Canby, Sheila R. 1997. "Islamic Lusterware." In Ian Freestone and David Gaimster (eds), *Pottery in the Making: World Ceramic Traditions*. London: British Museum Press.

Carusi, Paola. 2000. "Alchimia Islamica e Religione: la legittimazione difficile di una scienza della natura." In Carmela Baffioni (ed.), *Religion versus Science in Islam: A Medieval and Modern Debate*. Rome: Istituto per l'Oriente Carlo Alfonso Nallino.

Carusi, Paola. 2002. "Il trattato di filosofia alchemica 'Miftāḥ al-ḥikma' ed i suoi testimoni presso la Biblioteca Apostolica." *Miscellanea Bibliothecae Apostolicae Vaticanae*, 9: 35–84.

Carusi, Paola. 2003. "Il filosofo e il marinaio. Alchimia islamica e medicina alle prese con la natura." In Chiara Crisciani and Agostino Paravicini Bagliani (eds), *Alchimia e medicina nel Medioevo*. Florence: SISMEL – Edizioni del Galluzzo.

Carusi, Paola. 2005. "Alchimia Islamica e felicità nella *Risāla Ǧāmiʿa*: inalterabilità delle sostanze e pace dell'anima." In Maria Bettetini and Francesco D. Paparella (eds), *La Felicità nel Medioevo. Atti del convegno della società Italiana per lo studio del pensiero medievale (S.I.S.P.M.), Milano, 12–13 settembre 2003*. Louvain-la-Neuve: Brepols.

Celauro, Angela, David Loepp, and Daniela Ferro. 2017. "Ancient Procedures of Gold Cementation and Gold Scorification: Considerations on Their Reliability through Experimental Archaeology, Interpretation of Chemical Reactions and Thermodynamics." *Acta rerum naturalium*, 21: 177–200.

Chadwick, Adrian M., David R. Gilbert, and John Moore. 2012. "*... Quadrangles Where Wisdom Honours Herself*". *Archaeological Investigations at Tom Quad, Peckwater Quad and Blue Boar Quad, Christ Church, Oxford* (John Moore Heritage Services monograph, 1). Oxford: John Moore Heritage Services.

Chandelier, Joël. 2017. *Avicenne et la médecine en Italie. Le Canon dans les universités (1200–1350)*. Paris: Honoré Champion.

Châtillon, Jean (ed.). 1958. Richard of St Victor, *Liber exceptionum*. Paris: Vrin.

Cherry, John. 1991. "Leather." In John Blair and Nigel Ramsay (eds), *English Medieval Industries. Craftsmen, Techniques, Products*. London: Hambledon Press.

Colinet, Andrée. 2000a. "Le *Travail des quatre éléments* ou lorsqu'un alchimiste byzantin s'inspire de Jabir." In Isabelle Draelants, Anne Tihon, and Baudouin van den Abeele (eds), *Occident et Proche-Orient: Contacts scientifiques au temps des Croisades. Actes du colloque de Louvain-la-Neuve, 24 et 25 mars 1997*. Turnhout: Brepols.

Colinet, Andrée (ed.). 2000b. *Les Alchimistes grecs. T. X. L'Anonyme de Zuretti ou l'Art sacré et divin de la chrysopée par un anonyme*. Paris: Les Belles Lettres.

Contadini, Anna (ed.). 2007. *Arab Painting: Text and Image in Illustrated Manuscripts*. Leiden: Brill.

Corbin, Henry, and Pierre Lory. 1986. *L'Alchimie comme art hiératique* (Bibliothèque des mythes et des religions, 3). Paris: L'Herne.

Craddock, Paul T. 1995. *Early Metal Mining and Production*. Edinburgh: Edinburgh University Press.

Craddock, Paul T. (ed.). 1998. *2000 Years of Zinc and Brass* (Occasional Papers, 50). Rev. edn. London: British Museum.

Craddock, Paul T. 2003. "Cast Iron, Fined Iron, Crucible Steel: Liquid Iron in the Ancient World." In Paul T. Craddock and Janet Lang (eds), *Mining and Metal Production through the Ages*. London: British Museum.

Craddock, Paul T., Susan C. La Niece, and Duncan Hook. 1998. "Brass in the Medieval Islanmic World." In Paul T. Craddock (ed.), *2000 years of Zinc and Brass* (Occasional Papers, 50). Rev. edn. London: British Museum.

Crisciani, Chiara. 1973. "The Conception of Alchemy as Expressed in the *Pretiosa Margarita Novella* of Petrus Bonus of Ferrara." *Ambix*, 20: 165–81.

Crisciani, Chiara. 2002. *Il papa e l'alchimia. Felice V, Guglielmo Fabri e l'elixir*. Rome: Viella.

Crisciani, Chiara. 2006. "Tommaso, Pseudo-Tommaso e l'alchimia: Per un'indagine su un corpus alchemico." In Alessandro Ghisalberti, Antonio Petagine, and Rafaelle Rizzello (eds), *Letture e interpretazioni di Tommaso d'Aquino oggi: cantieri aperti. Atti del convegno internazionale di studio, Milano, 12–13 settembre 2005*. Torino: Istituto di filosofia S. Tommaso d'Aquino.

Crisciani, Chiara. Forthcoming. "Radical Moisture in the Alchemy of the Fourteenth and Fifteenth Centuries." In Jennifer M. Rampling and Peter M. Jones (eds), *Alchemy and Medicine from Antiquity to the Enlightenment*. London and New York: Routledge.

Crisciani, Chiara, and Giovanna Ferrari. 2010. "'Preface' and 'Commentary.'" In *Arnaldi de Villanova Opera medica omnia. Vol. V. Pars 2. Tractatus de humido radicali*. Barcelona: Pagès Ed.

Crisciani, Chiara, and Michela Pereira. 1998. "Black Death and Golden Remedies. Some Remarks on Alchemy and the Plague." In Agostino Paravicini Bagliani and Francesco Santi (eds), *The Regulation of Evil. Social and Cultural Attitudes to Epidemics in the Late Middle Ages*. Florence: SISMEL – Edizioni del Galluzzo.

Crisciani, Chiara, and Michela Pereira. 2008. "'Aurora consurgens': un dossier aperto." In *Natura, scienze e società medievali. Studi in onore di Agostino Paravicini Bagliani*. Florence: SISMEL – Edizioni del Galluzzo.

Crombie, Alistair Cameron. 1953. *Augustine to Galileo: The History of Science,* A.D. *400–1650*. Cambridge, MA: Harvard University Press.

Crossley, David W. 1967. "Glassmaking in Bagot's Park, Staffordshire in the 16th century." *Post-Medieval Archaeology*, 1: 44–83.

Dānish-Pazhūh, Muḥammad Taqī (ed.). 1964. Rāzī, Abū Bakr b. Zakariyyā'. *Kitāb al-asrār wa-sirr al-asrār*. Tehran: Commission Nationale Iranienne pour l'UNESCO.

Dapsens, Marion. 2016. "De la *Risālat Maryānus* au *De Compositione alchemiae*. Quelques réflexions sur la tradition d'un traité d'alchimie." *Studia Graeco-Arabica*, 6: 121–40.

Dapsens, Marion. Forthcoming. "The Alchemical Works Attributed to Khālid b. Yazīd." *Asiatische Studien*.

Dastin, John (pseud.?). 1702. "*Rosarium*." In Jean-Jacques Manget (ed.), *Bibliotheca chemica curiosa*. 2 vols. Geneva: Chouet.

Dautremont, Nathalie, et al. 2001. "La production potière des XIIIe–XVe siècles du quartier du Pontiffroy à Metz (Moselle): les fouilles de 1987–1988." *Revue archéologique de l'Est*, 51: 361–414.

De Juan Ares, Jorge, Noelia Fernández Calderón, Iván Muñiz López, Alejandro García Álvarez-Busto, and Nadine Schibille. 2018. "Islamic Soda-ash Glasses in the Christian Kingdoms of Asturias and León (Spain)." *Journal of Archaeological Science: Reports*, 22: 257–63.

De Smet, Daniel. 2003. "L'élaboration de l'élixir selon Ps. Siğistānī. Alchimie et cosmogonie dans l'ismaélisme ṭayyibite." In Alexander Fodor, Kinga Dévényi, and Tamás Iványi (eds), *Proceedings of the 20th congress of the Union Européenne des Arabisants et Islamisants (Part Two), Budapest, 10–17 September 2000*. Budapest: Csoma de Kőrös Soc.

Delva, Thijs. 2017. "The Abbasid Activist Ḥayyān al-ʿAṭṭār as the Father of Jābir b. Ḥayyān: An Influential Hypothesis Revisited." *Journal of Abbasid Studies*, 4: 35–61.

DeVun, Leah. 2009. *Prophecy, Alchemy, and the End of Time: John of Rupescissa in the Late Middle Ages*. New York: Columbia University Press.

Dinnetz, Mattias Karlsson. 2001. "Literary evidence for crucible steel in medieval Spain." *Historical Metallurgy*, 35(2): 74–80.

Dodge, Bayard. 1970. *The Fihrist of al-Nadīm. A Tenth-Century Survey of Muslim Culture* (Records of Civilization, Sources and Studies, 83). New York: Columbia University Press.

Dodwell, Charles Reginald (ed.). 1986. *Theophilus, The Various Arts, De Diversis Artibus*, 2nd ed. Oxford: Clarendon Press.

Dorveaux, Paul. 1913. *Le livre des simples médecines. Traduction française du Liber de simplici medicina dictus Circa instans de Platearius : tirée d'un manuscrit du XIIIe siècle (Ms. 3113 de la Bibliothèque Ste Geneviève de Paris)*. Paris: Société française d'histoire de la médecine.

Dufault, Olivier. 2015. "Transmutation Theory in the Greek Alchemical Corpus." *Ambix*, 62: 215–44.

Dumas, Geneviève. 2019. *Ymage de vie. Spéculation et expérimentation dans un traité d'alchimie médiévale*. Montpellier: Presses universitaires de la Méditerranée.

Dungworth, David. 2019. *Glassworking in England from the 14th to the 20th Century*. Swindon: Historic England.

Ercker, Lazarus. 1580. *Fleta minor. The Laws of Art and Nature in Knowing, Judging, Assaying, Refining and Inlarging Metals*. London: Thomas Dawks, 1683 (original German, 1580); see also Sisco and Smith 1951.

Evans, Allan (ed.). 1936. Pegolotti, Francesco Balducci. *La pratica della mercatura* (The Medieval Academy of America Publication, 24). Cambridge, MA: Medieval Academy of America.

Fauré, Benjamin. 2006. "Vers une histoire de l'alchimie médiévale en Occident, du domaine de la scolastique à celui du pouvoir temporel, avec une édition critique de la *Disputatio* attribuée à Michel Scot." Ph.D. thesis, University of Toulouse 2.

Ferrario, Gabriele. 2004. "Il Libro degli allumi e dei sali: *status quaestionis* e prospettive di studio." *Henoch*, 26: 275–96.

Ferrario, Gabriele. 2007. "Origins and Transmission of the *Liber de aluminibus et salibus*." In Lawrence M. Principe (ed.), *Chymists and Chymistry. Studies in the History of Alchemy and Early Modern Chemistry. Papers presented at an International Conference on the History of Alchemy and Chymistry, Held at the Chemical Heritage Foundation, Philadelphia, 19–23 July 2006*. Sagamore Beach, MA: Science History Publications.

Ferrario, Gabriele. 2010. "The Jews and Alchemy: Notes for a Problematic Approach." In Miguel López Pérez; Didier Kahn, and Mar Ray-Bueno (eds), *Chymia: Science and Nature in Early Modern Science (1450–1750)*. Newcastle: Cambridge Scholars Publishing.

Feuerbach, Ann M., David R. Griffiths, and John F. Merkel. 2003. "Early Islamic Crucible Steel Production at Merv, Turkmenistan." In Paul T. Craddock and Janrt Lang (eds), *Mining and Metal Production through the Ages*. London: British Museum.

Fidora, Alexander, and Dorothée Werner (eds). 2007. Gundissalinus, Dominicus. *De divisione philosophiae = Über die Einteilung der Philosophie: Lateinisch – Deutsch* (Herders Bibliothek der Philosophie des Mittelalters, 11). Freiburg: Herder.

Fierro, Maribel. 1996. "Bāṭinism in al-Andalus. Maslama b. Qāsim al-Qurṭubī (d. 353/964), Author of the *Rutbat al-Ḥakīm* and the *Ghāyat al-Ḥakīm (Picatrix)*." *Studia Islamica*, 84: 87–112.

Forbes, Robert J. 1970. *A Short History of the Art of Distillation: From the Beginnings up to the Death of Cellier Blumenthal*. Leiden: Brill.

Fors, Hjalmar, Lawrence M. Principe, and Heinz O. Sibum. 2016. "From the Library to the Laboratory and Back Again: Experiment as a Tool for Historians of Science." *Ambix*, 63: 85–97.

Forshaw, Peter J. 2013. "Cabala Chymica or Chemia Cabalistica – Early Modern Alchemists and Cabala." *Ambix*, 60: 361–89.

Forster, Regula. 2001. *Methoden mittelalterlicher arabischer Qur'ānexegese am Beispiel von Q 53, 1–18*. Berlin: Klaus Schwarz.

Forster, Regula. 2006. *Das Geheimnis der Geheimnisse: die arabischen und deutschen Fassungen des pseudo-aristotelischen Sirr al-asrār/Secretum secretorum* (Wissensliteratur im Mittelalter, 43). Wiesbaden: Reichert.

Forster, Regula. 2016. "Alchemy." In *Encyclopaedia of Islam. THREE*. Leiden: Brill.

Forster, Regula. 2017. *Wissensvermittlung im Gespräch. Eine Studie zu klassisch-arabischen Dialogen*. Leiden: Brill.

Forster, Regula, and Juliane Müller. 2020. "al-Jildakī." In *Encyclopaedia of Islam. THREE*. Leiden: Brill.

Franz, Marie-Louise von. 1957. *"Aurora consurgens", ein dem Thomas von Aquin zugeschriebenes Dokument der alchemistischen Gegensatzproblematik, herausgegeben und kommentiert (Carl Gustav Jung, Mysterium Coniunctionis, Teil 3)* (Psychologische Abhandlungen, 12). Zürich: Rascher.

Fratres Prædicatores (ed.). 1992. Thomas Aquinas. *Super Boethium de trinitate*. Rome: Commissio Leonina/Paris: Cerf.

Freestone, Ian. 1991. "Looking into glass." In Sheridan Bowman (ed.), *Science and the past*. London: British Museum Press.

Freestone, Ian Charles. 2006. "Glass Production in Late Antiquity and the Early Islamic Period: a Geochemical Perspective." In Marino Maggetti and Bruno Messiga (eds), *Geomaterials in Cultural Heritage* (Geological Society Special Publication, 257). Bath: Geological Society Publishing House.

Freestone, Ian Charles, Michael John Hughes, and Colleen P. Stapleton. 2008. "The Composition and Production of Anglo-Saxon Glass." In Vera Ivy Evison (ed.), *Catalogue of Anglo-Saxon glass in the British Museum*. London: British Museum.

Freestone, Ian Charles, Sophie Wolf, and Matthew Thirlwall. 2005. "The production of HIMT Glass: Elemental and Isotopic Evidence." In *Annales du 16e Congrès de l'Association Internationale pour l'Histoire du Verre*. Nottingham: AIHV.

Freudenthal, Gad. 1995. "Review of Raphael Patai, *The Jewish Alchemists: A History and Source Book*." *Isis*, 86: 318–19.

Freudenthal, Gad. 2011. "Alchemy in Medieval Jewish Cultures. A Noted Absence." In Gad Freudenthal (ed.), *Science in Medieval Jewish Cultures*. Cambridge: Cambridge University Press.

Gabriele, Mino. 1997. *Alchimia e Iconologia*. Udine: Forum.

Gabrovsky, Alexander N. 2015. *Chaucer the Alchemist*. New York: Palgrave Macmillan.

Ganzenmüller, Wilhelm. 1939. "Das Buch der Heiligen Dreifaltigkeit. Eine deutsche Alchemie aus dem Anfang des 15. Jahrhunderts." *Archiv für Kulturgeschichte*, 29: 93–146.

Ganzenmüller, Wilhelm. 1941. "*Liber florum Geberti*. Alchemistische Öfen und Geräte in einer Handschrift des 15. Jahrhunderts." *Quellen und Studien zur Geschichte der Naturwissenschaften und der Medizin*, 8: 273–303.

García Moreno, Renata, and Nicolas Thomas. 2008. "Cinnabar or Vermillion?" In Stefanos Kroustallis et al. (eds), *Art Technology. Sources and Methods. Proceedings of the Second Symposium of the Art Technological Source Research Working Group*. London: Archetype Publications.

Gayà Estelrich, Jordi (ed.). 1995. "Raimundi Lulli Liber de regionibus sanitatis et infirmitatis." In *Raimundi Lulli Opera latina, 20. [Opera] 106–113. In Monte Pessulano et Ianuae annis MCCIII–MCCCIV composita* (Corpus Christianorum, Continuatio Mediaevalis, 113). Turnhout: Brepols.

Gibb, Hamilton A.R. 1958. "Arab–Byzantine Relations under the Umayyad Caliphate." *Dumbarton Oaks Papers*, 12: 219–33.

Goitein, Shelomo Dov. 1967–1993. *A Mediterranean Society. The Jewish Communities of the Arab World as Portrayed in the Documents of the Cairo Geniza*. Berkeley: University of California Press.

Golb, Norman, Neḥumah B. Wahb, Ma'mar Ha-Sofer B. Isaac, and Moses B. Solomon. 1958. "Legal documents from the Cairo Genizah." *Jewish Social Studies*, 20: 17–46.

Gondonneau, Alexandra, and Maria Filomena Guerra. 2002. "The Circulation of Precious Metals in the Arab Empire: The Case of the Near and the Middle East." *Archaeometry*, 44: 573–99.

Gorzalczany, Amir, and Baruch Rosen. 2010. "A Possible Alchemist Apparatus from the Early Islamic period Excavated at Ramla, Israel." *Antiguo Oriente: Cuadernos del Centro de Estudios de Historia del Antiguo Oriente*, 8: 161–82.

Gowling, Eric. 2017. "Aetius' Extraction of Galenic Essence. A Comparison between Book 1 of Aetius' *Libri medicinales* and Galen's *On Simple Medicines*." In Lennart Lehmhaus and Matteo Martelli (eds), *Collecting Recipes. Byzantine and Jewish Pharmacology in Dialogue*. Boston, MA: De Gruyter.

Goy, Corinne. 1995. "Récipients en terre cuite d'un atelier de distillation fin XIVe–début XVe siècles." In *Ex pots: céramiques médiévales et modernes en Franche-Comté. Catalogue d'exposition, Musée des ducs de Wurtemberg, 1995*. Montbéliard: Musée des ducs de Wurtemberg.

Gratuze, Bernard, and Jean-Noël Barrandon. 1990. "Islamic Glass Weights and Stamps – Analysis using Nuclear Techniques." *Archaeometry*, 32: 155–62.

Greenaway, Frank. 1972. "Introduction In Stephen Moorhouse, Frank Greenaway, Charles C. Moore, C. Vincent Bellamy, W. E. Nicolson, and Leo Biek. 1972. '*Medieval Distilling Apparatus of Glass and Pottery*.'" *Medieval Archaeology*, 16: 79–121.

Guadagnin, Rémy. 2007. *Fosses. Vallée de l'Ysieux. Mille ans de production céramique en Île-de-France. Volume 2: Catalogue typo-chronologique des productions*. Caen: Publications du CRAHM.

Günther, Sebastian. 2017. "Education, general (up to 1500)." In *Encyclopaedia of Islam. THREE*. Leiden: Brill.

Gutas, Dimitri. 1998. *Greek Thought, Arabic Culture: The Graeco-Arabic Translation Movement in Baghdad and Early 'Abbāsid Society (2nd–4th = 8th–10th centuries)*. London: Routledge.

Haldane, Duncan. 1978. *Mamluk Painting*. Warminster: Aris and Philips.

Hall, Bert. 1996. "The Corning of Gunpowder and the Development of Firearms in the Renaissance." In Brenda J. Buchanan (ed.), *Gunpowder: The History of an International Technology*. Bath: Bath University Press.

Halleux, Robert. 1973. *Le problème des métaux dans la science antique* (Bibliothèque de la faculté de philosophie et lettres de l'Université de Liège, 209). Liège: Presses universitaires de Liège.

Halleux, Robert. 1979. *Les textes alchimiques* (Typologie des sources du Moyen Âge occidental, 32). Turnhout: Brepols.
Halleux, Robert. 1981. "Les ouvrages alchimiques de Jean de Rupescissa." In *Histoire littéraire de la France. Tome XLI, suite du quatorzième siècle*. Paris: Imprimerie nationale.
Halleux, Robert. 1982. "Albert le Grand et l'alchimie." *Revue des sciences philosophiques et théologiques*, 66: 57–80.
Halleux, Robert, and Paul Meyvaert. 1987. "Les origines de la *Mappae clavicula*." *Archives d'Histoire Doctrinale et Littéraire du Moyen Âge*, 54: 7–58.
Hallum, Benjamin. 2008. *Zosimus Arabus: The Reception of Zosimos of Panopolis in the Arabic/Islamic World*. Ph.D. thesis, Warburg Institute, London.
Hallum, Benjamin. 2009. "The *Tome of Images*: An Arabic Compilation of Texts by Zosimos of Panopolis and a Source of the *Turba Philosophorum*." *Ambix*, 56: 76–88.
Hallum, Benjamin, and Marcel Marée. 2016. "A Medieval Alchemical Book Reveals New Secrets." *The British Museum Blog*. Available online: https://blog.britishmuseum.org/a-medieval-alchemical-book-reveals-new-secrets/ (accessed August 25, 2021).
Haq, Syed Nomanul. 1994. *Names, Natures and Things, the Alchemist Jābir ibn Ḥayyān and His Kitāb al-Aḥjār (Book of Stones)* (Boston Studies in the Philosophy of Science, 158). Dordrecht: Kluwer.
Harig, Georg. 1974. *Bestimmung der Intensität im medizinischen System Galens. Ein Beitrag zur theoretischen Pharmakologie, Nosologie und Therapie in der Galenischen Medizin*. Berlin: Akademie-Verlag.
Harris, Nicholas G. 2017. "In Search of 'Izz al-Dīn Aydamir al-Ǧildakī, Mamlūk Alchemist." *Arabica*, 64: 531–56.
al-Hassan, Ahmad Y. 2004. "The Arabic Original of *Liber de compositione alchemiae*, the Epistle of Maryānus, the Hermit and Philosopher, to Prince Khālid ibn Yazīd." *Arabic Sciences and Philosophy*, 14: 213–31.
al-Hassan, Ahmad Y., and Donald R. Hill. 1986. *Islamic Technology: An Illustrated History*. Cambridge: Cambridge University Press.
Hawthorne, John G., and Cyril Stanley Smith. 1979. *Theophilus. On Divers Arts: The Foremost Medieval Treatise on Painting, Glassmaking and Metalwork*. New York: Dover.
al-Ḥaẓīmī, Nāṣir. 2003. *Ḥarq al-kutub fī l-turāth al-ʿarabī*. Cologne: Manshūrāt al-Jamal.
Hein, Christel. 1985. *Definition und Einteilung der Philosophie von der spätantiken Einleitungsliteratur zur arabischen Enzyklopädie*. Frankfurt: Peter Lang.
Hellwig, Oliver. 2009. *Wörterbuch der mittelalterlichen indischen Alchemie*. Havertown, PA: Barkhuis Publishing.
Hernández Sobrino, Angel M. 1996. "Hornos medievales de azogue en Almadén." In *Actas de las I Jornadas sobre Mineraria y Tecnología en la Edad Media Peninsular (Léon, 26–29 septembre 1995)*. Leon: Fundación Hullera Vasco-Leonesa.
Hill, C.R. 1975. "The Iconography of the Laboratory." *Ambix*, 22: 102–10.
Hinckley, Marlis Ann. 2017. "Diagrams and Visual Reasoning in Pseudo-Lullian Alchemy, 1350–1500." Master of Studies thesis, King's College, University of Cambridge.
Hirschler, Konrad. 2012. *The Written Word in the Medieval Arabic Lands. A Social and Cultural History of Reading Practices*. Edinburgh: Edinburgh University Press.

Holl, Imre. 1982. "Középkori desztilláló készülékek cserépből Kőszeg várában." *Archaeologiai Értesítő*, 109: 108–23.

Holmyard, Eric John (ed.). 1923. Abū al-Qāsim al-ʿIrāqī. *Kitāb al-ʿilm al-muktasab fī zirāʿat al-dhahab*. Paris: Geuthner.

Holmyard, Eric John. 1924. "Maslama al-Majrīṭī and the *Rutbatu'l-Ḥakīm*." *Isis*, 6: 293–305.

Holmyard, Eric John. 1926. "Abu' l-Qāsim al-ʿIrāqī." *Isis*, 8: 403–26.

Holmyard, Eric John. 1928. *The Arabic Works of Jâbir ibn Ḥayyân edited/Muṣannafāt fī ʿilm al-kīmiyāʾ li-l-ḥakīm Jābir ibn Ḥayyān al-Ṣūfī*. Paris: Geuthner.

Holmyard, Eric John. 1937. "Aidamir al-Jildakī." *Iraq*, 4: 47–53.

Holmyard, Eric John. 1956. "Alchemical Equipment." In Charles Singer, Eric John Holmyard, A. Rupert Hall, and Trevor I. Williams (eds), *A History of Technology. Vol. II. The Mediterranean Civilizations and the Middle Ages, c. 700 B.C. to c. A.D. 1500*. Oxford: Clarendon Press.

Holmyard, Eric John, and Desmond Christopher Mandeville. 1927. *Avicennae De congelatione et conglutinatione lapidum: Being Sections of the Kitâb al-shifâ'*. Paris: Geuthner.

Hoover, Herbert Clark, and Lou Henry Hoover. 1950. *Georgius Agricola's De Re Metallica*. New York: Dover Publications.

Horgan, Frances (transl.). 1994. *The Romance of the Rose*. Oxford: Oxford University Press.

Hoyland, Robert G., and Brian Jeremy James Gilmour. 2006. *Medieval Islamic Swords and Swordmaking: Kindī's Treatise "On Swords and their Kinds."* Warminster: Gibb Memorial Trust.

Hudry, Françoise. 1997–1999. "Le *De secretis nature* du Ps. Apollonius de Tyane, traduction latine par Hugues de Santalla du *Kitāb sirr al-khalīqa*." *Chrysopœia*, 6: 1–154.

Ibn al-Ḥājj, Muḥammad b. Muḥammad al-Fāsī. n.d. *al-Mudkhal*. 4 vols. Cairo: Maktabat Dār al-turāth.

Ibn Farīghūn. 1985. *Compendium of sciences = Jawāmiʿ al-ʿulūm*. Frankfurt a. M.: Institute for the History of Arabic-Islamic Science.

Ibn Ḥajar al-ʿAsqalānī. n.d. *al-Durar al-kāmina*, vol. III. Beirut: Dār al-Jīl.

Ibn Sīnā. n.d. "Risāla fī aqsām al-ʿulūm al-ʿaqliyya." In Ibn Sīnā, *Tisʿ rasāʾil fī l-ḥikma wa-l-ṭabīʿiyyāt*. Cairo: Dār al-ʿarab.

Ilg, Albert. 1874. *Theophilus presbyter schedula diversarum artium: I. Band* (Quellenschriften für Kunstgeschichte und Kunsttechnik des Mittelalters und der Renaissance, 7). Vienna: Wilhelm Braumüller.

Institut du monde arabe. 1996. *À l'ombre d'Avicenne. La médecine au temps des califes. Exposition présentée du 18 novembre 1996 au 2 mars 1997*. Paris: Snoeck-Ducaju & Zoon, Institut du monde arabe.

Iskandar, A.Z. 1984. *A Descriptive List of Arabic Manuscripts on Medicine and Science at the University of California, Los Angeles*. Leiden: Brill.

Jacoby, David. 1993. "Raw Materials for the Glass Industries of Venice and the Terraferma about 1370–about 1460." *Journal of Glass Studies*, 35: 65–90.

James, Liz 2006. "Byzantine Glass Mosaic Tesserae: Some Material Considerations." *Byzantine and Modern Greek Studies*, 30: 29–47.

Jennemann, Hans R. 1997. *Die Waage des Chemikers. The Chemist's Balance*. Frankfurt: DECHEMA.

Jockenhövel, Albrecht (ed.). 2013. *Mittelalterliche Eisengewinnung im Märkischen Sauerland. Archäometallurgische Untersuchungen zu den Anfängen der Hochofentechnologie in Europa [Medieval Iron Production in the Märkisches Sauerland]*. Rahden: Verlag Marie Leidorf.

Johnson, Kathryn V. 1996. "Jalāl al-Dīn Rūmī's Use of Alchemical Imagery." *Islamic Culture*, 70: 1–25.

Johnson, Rozelle Parker. 1939. *Compositiones Variae, from Codex 490, Biblioteca Capitolare, Lucca, Italy, an Introductory Study* (Illinois Studies in Language and Literature, 23/3). Urbana: University of Illinois Press.

Joose, N. Peter. 2008. "Unmasking the Craft: 'Abd al-Laṭīf al-Baghdādī's Views on Alchemy and Alchemists." In Anna Akasoy and Wim Raven (eds), *Islamic Thought in the Middle Ages: Studies in Text, Transmission and Translation in Honour of Hans Daiber*. Leiden: Brill.

Jouttijärvi, Arne, Turi Thomsen, and Annine S.A. Moltesen. 2005. "Værkstedets function [The metal workshop]." In Mette Iversen, David Earle Robinson, Jesper Hjermind, and Charlie Christensen (eds), *Viborg Søndersø 1018–1030. Arkæologi og naturvidenskab i et værkstedsområde fra vikingetid* (Skrifter, 52). Højbjerg: Jysk Arkæologisk Selskabs.

Judy, Albert G. (ed.). 1976. Robert Kilwardby, *De ortu scientiarum*. Toronto: Pontifical Institute of Mediaeval Studies.

Kahn, Didier. 1990–1. "Note sur deux manuscrits du Prologue attribué à Robert de Chester." *Chrysopœia*, 4: 33–4.

Kahn, Didier. 2010a. "Alchemical Poetry in Medieval and Early Modern Europe: A Preliminary Survey and Synthesis. Part I – Preliminary Survey." *Ambix*, 57: 249–74.

Kahn, Didier. 2010b. "The *Turba philosophorum* and its French Version (15th c.)." In Miguel López Pérez, Didier Kahn, and Mar Rey-Bueno (eds), *Chymia: Science and Nature in Early Modern Science (1450–1750)*. Newcastle: Cambridge Scholars Publishing.

Kahn, Didier. 2011. "Alchemical Poetry in Medieval and Early Modern Europe: A Preliminary Survey and Synthesis. Part II – Synthesis." *Ambix*, 58: 62–77.

Kahn, Didier. 2017. "Généalogie de l'alchimie et interprétation alchimique de la Bible au XIVe siècle: *Qui fuerint primi inventores hujus artis*." *Archives d'Histoire Doctrinale et Littéraire du Moyen Âge*, 84: 313–47.

Kamāl, Aḥmad Bak (ed.). 1907. *Kitāb al-Durr al-maknūz wa-l-sirr al-ma'zūz fī l-dalā'il wa-l-khabāyā wa-dafā'in wa-l-kunūz*. 2 vols. Cairo: Institut français d'archéologie orientale.

Kamber, Pia, and Peter Kurzmann. 2002. "Ein metallurgisches Laboratorium des 13. Jahrhunderts in Basel." In Guido Helmig, Barbara Scholkmann, and Matthias Untermann (eds), *Medieval Europe Basel 2002. 3. Internationaler Kongress der Archäologie des Mittelalters und der Neuzeit vom 10. bis 15. September 2002 in Basel*. Basel: Archaologische Bodenforschung Basel-Stadt.

Kamber, Pia, Peter Kurzmann, and Gerber Yvonne. 1999. "Der Gelbschmied und Alchemist (?) von Ringelhof." *Jahresbericht der Archäologischen Bodenforschung des Kantons Basel-Stad*, 1998: 151–99.

Karimi Zanjani Asl, Mohammad. 1391 SH/2013 AD. "al-Aṣnām al-sab'a." In *Danishnāma-yi farhang-i mardum-i Īrān (The Encylopaedia of Iranian Folklore), Vol. I*. Tehran: Markaz-i Dā'irat ul-ma'arif-i buzurg-i islāmī (The Centre for the Great Islamic Encyclopaedia).

Käs, Fabian. 2010. *Die Mineralien in der arabischen Pharmakognosie. Eine Konkordanz zur mineralischen Materia medica der klassischen arabischen Heilmittelkunde nebst überlieferungsgeschichtlichen Studien* (Akademie der Wissenschaften und der Literatur. Veröffentlichungen der orientalischen Kommission, 54). 2 vols. Wiesbaden: Harrassowitz.

Kenward, Harry K., and Allan R. Hall, 1995. *Biological Evidence from Anglo-Scandinavian Deposits at 16–22 Coppergate* (Archaeology of York, 14/7). York: Council for British Archaeology.

Kibre, Pearl. 1942. "Alchemical Writings Attributed to Albertus Magnus." *Speculum*, 17: 499–518.

Kibre, Pearl. 1944. "An Alchemical Tract Attributed to Albertus Magnus." *Isis*, 35: 303–16.

King, Peter. 2016. "The Zenith of Iron and the Transition to Mild Steel in Great Britain." *Historical Metallurgy*, 50: 109–22.

Kiss, Farkas Gábor, Benedek Láng, and Cosmin Popa-Gorjanu. 2006. "The Alchemical Mass of Nicolaus Melchior Cibinensis: Text, Identity and Speculations." *Ambix*, 53: 143–59.

Kohler, Robert E. 2008. "Lab History: Reflections." *Isis*, 99: 761–8.

Kouhkan, Reza. 2015. *Pensée alchimique de Tughrâï*. Saarbrücken: Editions Universitaires Européennes.

Kraus, Paul. 1935. *Jābir ibn Ḥayyān, Essai sur l'histoire des idées scientifiques dans l'Islam. Vol. I: Textes choisis/Mukhtār rasā'il Jābir ibn Ḥayyān*. Paris: Maisonneuve.

Kraus, Paul. 1942. *Jābir ibn Ḥayyān, contribution à l'histoire des idées scientifiques dans l'Islam. Vol. II: Jābir et la science grecque* (Mémoires présentés à l'Institut d'Égypte, 45). Cairo: Institut français d'archéologie orientale.

Kraus, Paul. 1943. *Jābir ibn Ḥayyān, contribution à l'histoire des idées scientifiques dans l'Islam. Vol. I: Le corpus des écrits jābiriens* (Mémoires présentés à l'Institut d'Égypte, 44). Cairo: Institut français d'archéologie orientale.

Krueger, Ingeborg. 2014. "Islamisches Bleiglas und *mīnā*." *Journal of Glass Studies*, 56: 61–84.

Kurzmann, Peter. 2000. *Die Destillation im Mittelalter: archäologische Funde und Alchemie*. Schloss Hohentübingen: Verl. des Vereins für Archäologie des Mittelalters.

Kurzmann, Peter. 2009. "Einige Glasgeräte der arabischen Alchemie." *Sudhoffs Archiv*, 93(2): 184–200.

L'Escalopier, Charles de (ed.). 1843. *Théophile, prêtre et moine, Essai sur divers arts, Diversarum artium schedula*. Paris: Firmin Didot Frères.

L'Heritier, Maxime, and Florian Téreygeol. 2010. "From Copper to Silver: Understanding the *Saigerprozess* through Experimental Liquation and Drying." *Historical Metallurgy*, 44: 136–52.

Lacaze, Grégoire. 2018. *Turba philosophorum. Congrès pythagoricien sur l'art d'Hermès* (Philosophia antiqua, 150). Leiden: Brill.

Lamm, Kristina. 2008. "Crucibles and cupels from building group 3." In Helen Clarke and Kristina Lamm (eds), *Excavations at Helgö XVII. Workshop, Part III*. Stockholm: Kungl. Vitterhets Historie och Antikvitets Akademien.

Langermann, Y. Tzvi. 1996. "Review of Raphael Patai, *The Jewish Alchemists: A History and Source Book*." *Journal of the American Oriental Society*, 116: 792–3.

Leenhardt, Marie. 1995. "Céramiques, métal et alchimie." In *Poteries d'Oc: céramiques languedociennes, VIIe–XVIIe siècles (catalogue de l'exposition au Musée archéologique de Nîmes, 15 octobre 1995–28 février 1996)*. Aix-en-Provence: Narration.

Lemay, Richard. 1990–1991. "L'authenticité de la Préface de Robert de Chester à sa traduction du *Morienus*." *Chrysopœia*, 4: 3–32.

Lennep, Jacques van. 1984. *Alchimie. Contribution à l'histoire de l'art alchimique*. Bruxelles: Crédit Communal.

Letrouit, Jean. 1995. "Chronologie des alchimistes grecs." In Didier Kahn and Sylvain Matton (eds), *Alchimie, art, histoire et mythes: actes du 1er Colloque internationale de la Société d'étude de l'histoire de l'alchimie (Paris, Collège de France, 14–15–16 mars 1991)*. Paris: S.É.H.A.-ARCHÈ.

Leube, Georg. 2013. *Die Rezepte der Freiburger alchemistischen Handschrift des 'Abd al-Ǧabbār al-Hamaḏānī: Edition, Übersetzung und Kommentar* (Islamkundliche Untersuchungen, 315). Berlin: Klaus Schwarz Verlag.

Levey, Martin. 1955. "Some Chemical Apparatus of Ancient Mesopotamia." *Journal of Chemical Education*, 32: 180–3.

Liebrenz, Boris. 2016. *Die Rifā'īya aus Damaskus. Eine Privatbibliothek im osmanischen Syrien und ihr kulturelles Umfeld*. Leiden: Brill.

Lippmann, Edmund Oskar von. 1919. *Entstehung und Ausbreitung der Alchemie*. 3 vols. Berlin: Springer.

Livingston, John W. 1971. "Ibn Qayyim al-Jawziyyah: A Fourteenth Century Defense against Astrological Divination and Alchemical Transmutation." *Journal of the American Oriental Society*, 91(1): 96–103.

Lory, Pierre. 2016. "Aspects de l'ésotérisme chiite dans le Corpus Ǧābirien: Les trois Livres de l'Élément de fondation." *Al-Qanṭara*, 37: 279–98.

Luanco, José Ramón de. 1980. *La Alquimia en Espana*. Madrid: Editorial "Tres, Catorce, Diecisiete."

Lubac, Henri de. 1959–1964. *Exégèse médiévale: les quatre sens de l'Écriture*. 4 vols. Paris: Aubier.

Lüthy, Christoph, and Alexis Smets. 2009. "Words, Lines, Diagrams, Images: Towards a History of Scientific Imagery." *Early Science and Medicine*, 14: 398–439.

Madelung, Wilferd. 2014–2015. "Maslama al-Qurṭubī's Contribution to the Shaping of the Encyclopedia of the Ikhwān al-Ṣafā'." In Antonella Straface, Carlo de Angelo, and Andrea Manzo (eds), *Labor limae. Atti in onore di Carmela Baffioni*. Naples: Centro di Studi Magrebini.

Madelung, Wilferd. 2017a. "Maslama al-Qurṭubī's *Kitāb Rutbat al-ḥakīm* and the History of Chemistry." *Intellectual History of the Islamicate World*, 5: 118–26.

Madelung, Wilferd. 2017b. *The Book of the Rank of the Sage, Rutbat al-Ḥakīm by Maslama b. Qāsim al-Qurṭubī. Arabic Text with an Introduction* (Corpus Alchemicum Arabicum, 4). Zürich: Living Human Heritage Publications.

Madkūr, Ibrāhīm et al. (eds). 1964. Ibn Sīnā. *Al-Shifā'. Al-Ṭabī'iyyāt. 5, al-ma'ādin wa-l-āthār al-'ulwiyya (La physique. 5, Les métaux et la météorologie)*. Cairo: al-Hay'a al-miṣriyya al-'āmma li-l-kitāb.

Magnusson, Gert (ed.). 1995. *The Importance of Ironmaking: Technical Innovation and Social Change. Papers Presented at the Norberg Conference 1995, Vol. I.* Stockholm: Jernkontoret.

Maire, Jean, and Jean-Pierre Rieb. 1972. "Un puits du XVe siècle dans le Marais-Vert à Strasbourg." *Cahiers alsaciens d'archéologie, d'art et d'histoire*, 16: 165–80.

Mandosio, Jean-Marc. 1991–1993. "La place de l'alchimie dans les classifications des sciences et des arts à la Renaissance." *Chrysopœia*, 4: 199–282.

Mandosio, Jean-Marc. 1993. "L'alchimie dans les classifications des sciences et des arts à la Renaissance." In Jean-Claude Margolin and Sylvain Matton (eds), *Alchimie et philosophie à la Renaissance. Actes du colloque international de Tours, 1991*. Paris: Vrin.

Mandosio, Jean-Marc. 2000. "Commentaire alchimique et commentaire philosophique." In Marie-Odile Goulet-Cazé (ed.), *Le Commentaire, entre tradition et innovation. Actes du Colloque international de l'Institut des traditions textuelles (Paris et Villejuif, 22–25 septembre 1999)*. Paris: Vrin.

Mandosio, Jean-Marc. 2001. "Les lexiques bilingues philosophiques, scientifiques et notamment alchimiques à la Renaissance." In Jacqueline Hamesse and Danielle Jacquart (eds), *Lexiques bilingues dans les domaines philosophique et scientifique (Moyen Âge – Renaissance)*. Turnhout: Brepols.

Mandosio, Jean-Marc. 2004. "La *Tabula smaragdina* e i suoi commentari medievali." In Paolo Lucentini, Ilaria Parri, and Vittoria Perrone Compagni (eds), *Hermetism from Late Antiquity to Humanism. La tradizione ermetica dal mondo tardo-antico all'umanesimo. Atti del Convegno internazionale di studi, Napoli, 20–24 novembre 2001*. Turnhout: Brepols.

Mandosio, Jean-Marc. 2005. "La création verbale dans l'alchimie latine du Moyen Âge." *Archivum Latinitatis Medii Aevi*, 63: 137–47.

Mandosio, Jean-Marc. 2012. "Basilisks, Lettuce, and the Stone Which Is Not a Stone: On the Relationship between Living Things and Inert Substances in Medieval Alchemy." In Danielle Jacquart and Nicolas Weill-Parot (eds), *Substances minérales et corps animés, Mineral Substances and Animate Bodies, 1100–1500*. Montreuil: Omniscience.

Mandosio, Jean-Marc. 2017. "La classification des sciences dans le *Miroir des amoureux* et l'érotologie de Jean de Meun." In Jean-Patrice Boudet, Philippe Haugeard, Silvère Menegaldo, and François Ploton-Nicollet (eds), *Jean de Meun et la culture médiévale: littérature, art, sciences et droit aux derniers siècles du Moyen Âge*. Rennes: Presses universitaires de Rennes.

Mandosio, Jean-Marc. 2018a. "Follower or Opponent of Aristotle? The Critical Reception of Avicenna's Meteorology in the Latin World and the Legacy of Alfred the Englishman." In Dag N. Hasse and Amos Bertolacci (eds), *The Arabic, Hebrew and Latin Reception of Avicenna's Physics and Cosmology*. Berlin: De Gruyter.

Mandosio, Jean-Marc (ed.). 2018b. Lefèvre d'Etaples, Jacques. *La Magie naturelle = De magia naturali, vol. 1: L'influence des astres* (Bibliothèque secrète, 1). Paris: Les Belles Lettres.

Mandosio, Jean-Marc. 2019. "Peter of Zealand." In Sophie Page and Catherine Rider (eds), *The Routledge History of Medieval Magic*. Abingdon: Routledge.

Mandosio, Jean-Marc, and Carla Di Martino. 2006. "La 'Météorologie' d'Avicenne (*Kitāb al-Shifā' V*) et sa diffusion dans le monde latin." In *Wissen über Grenzen: arabisches Wissen und lateinisches Mittelalter*. Berlin: De Gruyter.

Manget, Jean-Jacques (ed.). 1702. *Bibliotheca Chemica Curiosa*. Geneva: Chouet.

Maqbūl, Aḥmad. 1929. "A Persian Translation of the Eleventh Century Arabic Alchemical Treatise *'Ain aṣ-ṣan'ah wa 'Aun aṣ-ṣan'ah*." *Memoirs of the Asiatic Society of Bengal*, 8(7): 419–60.

Margoliouth, David S., and Eric John Holmyard. 1931. "Arabic Documents from the Monneret Collection." *Islamica*, 4: 249–71.

Marinovic-Vogg, Marianne. 1990. "'Son of Heaven.' The Middle Netherlands Translation of the Latin Tabula Chemica." In Zweder R.W.M. von Martels (ed.), *Alchemy Revisited. Proceedings of the International Conference on the History of Alchemy at the University of Groningen, 17–19 April 1989*. Leiden: Brill.

Martelli, Matteo. 2009. "'Divine Water' in the Alchemical Writings of Pseudo-Democritus." *Ambix*, 56: 5–22.
Martelli, Matteo. 2011. "Greek Alchemists at Work: 'Alchemical Laboratory' in the Greco-Roman Egypt." *Nuncius*, 26: 271–311.
Martelli, Matteo. 2013. *The Four Books of Pseudo-Democritus* (Sources of alchemy and chemistry, 1). Leeds: Maney.
Martelli, Matteo. 2014a. "L'alchimie en syriaque et l'œuvre de Zosime." In Émilie Villey (ed.), *Les sciences en syriaque. Actes de la 11e Table ronde de la Société d'études syriaques, Paris, 15 novembre 2013*. Paris: Geuthner.
Martelli, Matteo. 2014b. "Properties and Classification of Mercury between Natural Philosophy, Medicine, and Alchemy." *AION, Annali dell'Università degli Studi di Napoli "L'Orientale"*, 36: 17–47.
Martelli, Matteo. 2017. "Translating Ancient Alchemy: Fragments of Graeco-Egyptian Alchemy in Arabic Compendia." *Ambix*, 64: 1–17.
Martinón-Torres, Marcos. 2011. "Some Recent Developments in the Historiography of Alchemy." *Ambix*, 58: 215–37.
Mathis, François, Olivier Vrielynck, Kilian Laclavetine, Grégoire Chêne, and David Strivay. 2008. "Study of the provenance of Belgian Merovingian garnets by PIXE at IPNAS cyclotron." *Nuclear Instruments and Methods in Physics Research B*, 266: 2348–52.
Matton, Sylvain. 2009. *Scolastique et alchimie (XVIe–XVIIe siècles)* (Philosophie et alchimie à la Renaissance et à l'Âge classique, 1). Paris: S.É.H.A.-ARCHÈ.
Matus, Zachary. 2013. "Resurrected Bodies and Roger Bacon's Elixir." *Ambix*, 60: 323–40.
McCallum, R. Ian. 2007. "Alchemical Scrolls Associated with George Ripley." In Stanton J. Linden (ed.), *Mystical Metal of Gold: Essays on Alchemy and Renaissance Culture*. New York: AMS.
Mehren, August Ferdinand Michael van, and Christian Martin Fraehn (eds). 1866. Shams al-dīn al-Dimashqī. *Cosmographie de Chems-ed-din Abou Abdallah Mohammed ed-Dimichqui/Kitāb nukhbat al-dahr fī 'ajā'ib al-barr wa-l-baḥr*. Saint Petersburg: Académie impériale des sciences.
Meller, Harald, Alfred Reichenberger, and Christian-Heinrich Wunderlich (eds). 2016. *Alchemie und Wissenschaft des 16. Jahrhunderts. Fallstudien aus Wittenberg und vergleichbare Befunde: internationale Tagung vom 3. bis 4. Juli 2015 in Halle (Saale)* (Tagungen des Landesmuseums für Vorgeschichte Halle, 15). Halle: Landesamt für Denkmalpflege und Archäologie Sachsen-Anhalt Landesmuseum für Vorgeschichte.
Mentgen, Gerd. 2009. "Jewish Alchemists in Central Europe in the Later Middle Ages: Some New Sources." *Aleph*, 9: 345–52.
Merkel, Stephen W., Leonid Sverchkov, Andreas Hauptmann, Volker Hilberg, Martin Bode, and Robert Lehmann. 2013. "Analysis of Slag, Ore, and Silver from the Tashkent and Samarkand Areas: Medieval Silver Production and the Coinage of Samanid Central Asia." In Andreas Hauptmann, Oliver Mecking, and Matthias Prange (eds), *Archäometrie und Denkmalpflege 2013* (Metalla Sonderhefte, 6). Bochum: Deutsches Bergbau-Museum.
Merrified, Mary P. 1967 [1st ed. 1849]. *Original Treatises on the Arts of Painting*. Reprint. New York: Dover.
Mertens, Michèle (ed.). 1995. Zosime de Panopolis. *Les Alchimistes grecs. T. IV, 1. Zosime de Panopolis. Mémoires authentiques*. Paris: Les Belles Lettres.

Messier, Ronald A. 1974. "The Almoravids: West African Gold and the Gold Currency of the Mediterranean Basin." *Journal of the Economic and Social History of the Orient*, 17: 31–42.

Miller, Pat, and Roy Stephenson. 1999. *A 14th-Century Pottery Site in Kingston upon Thames, Surrey, Excavations at 70–76 Eden Street* (MoLAS archaeology studies series, 1). London: Museum of London Archaeology Service.

Moorhouse, Stephen. 1981. "The Medieval Pottery Industry and Its Markets." In David W. Crossley (ed.), *Medieval Industry*. London: Council for British Archaeology.

Moorhouse, Stephen. 1983. "Pottery and Vessel Glass." In Philip Mayes, Lawrence A. S. Butler, and Shirley Johnson (eds), *Sandal Castle excavations 1964–1973. A Detailed Archaeological Report*. Wakefield: Wakefield Historical.

Moorhouse, Stephen, Frank Greenaway, Charles C. Moore, C. Vincent Bellamy, W.E. Nicolson, and Leo Biek. 1972. "Medieval Distilling-Apparatus of Glass and Pottery." *Medieval Archaeology*, 16: 79–121.

Moureau, Sébastien. 2012. "Les sources alchimiques de Vincent de Beauvais." *Spicae, Cahiers de l'Atelier Vincent de Beauvais, nouvelle série*, 2: 5–118. Available online: http://spicae-cahiers.irht.cnrs.fr/content/les-sources-alchimiques-de-vincent-de-beauvais (accessed August 25, 2021).

Moureau, Sébastien. 2013. "*Elixir Atque Fermentum*. New Investigations about the Link between Pseudo-Avicenna's Alchemical *De anima* and Roger Bacon: Alchemical and Medical Doctrines." *Traditio, Studies in Ancient and Medieval Thought, History, and Religion*, 68: 277–323.

Moureau, Sébastien. 2016a. *Le* De anima *alchimique du pseudo-Avicenne. Vol. 1. Étude*. Florence: SISMEL – Edizioni del Galluzzo.

Moureau, Sébastien. 2016b. *Le* De anima *alchimique du pseudo-Avicenne. Vol. 2. Édition critique et traduction annotée*. Florence: SISMEL – Edizioni del Galluzzo.

Moureau, Sébastien. 2020. "*Min al-Kīmiyā' ad Alchimiam*. The Transmission of Alchemy from the Arab-Muslim world to the Latin West in the Middle Ages." *Micrologus*, 28: 87–141.

Moureau, Sébastien. Forthcoming. "Alchemical Equipment." In Sonja Brentjes (ed.), *Routledge Handbook on Science in the Islamicate World: Practices from the 8th to the 19th Century*. Abingdon: Routledge.

Moureau, Sébastien, and Nicolas Thomas. 2015. "L'aludel: savoir et savoir-faire transmis du monde arabe à l'Occident médiéval?" In Roland-Pierre Gayraud, Jean-Michel Poisson, and Catherine Richarté (eds), *Héritages arabo-islamiques dans l'Europe méditerranéenne*. Paris: La Découverte.

Moureau, Sébastien, and Nicolas Thomas. 2016. "Understanding Texts with the Help of Experimentation: The Example of Cupellation in Arabic Scientific Literature." *Ambix*, 63: 98–117.

Müller, Juliane. 2012. *Zwei arabische Dialoge zur Alchemie: die Unterredung des Aristoteles mit dem Inder Yūhīn und das Lehrgespräch der Alchemisten Qaydarūs und Mītāwus mit dem König Marqūnus: Edition, Übersetzung und Kommentar* (Islamkundliche Untersuchungen, 310). Berlin: Klaus Schwarz.

Neven, Sylvie. 2014. "Transmission of Alchemical and Artistic Knowledge in German Mediaeval and Premodern Recipe Books." In Sven Dupré (ed.), *Laboratories of Art. Alchemy and Art Technology from Antiquity to the 18th Century*. Cham: Springer.

Newman, William R. 1986. "The *Summa perfectionis* and Late Medieval Alchemy. A Study of Chemical Traditions, Techniques and Theories in Thirteenth Century Italy." Ph.D. thesis, Harvard University, Cambridge, MA.

Newman, William R. 1989. "Technology and Alchemical Debate in the Late Middle Ages." *Isis*, 80: 423–45.
Newman, William R. 1991. *The Summa Perfectionis of Pseudo-Geber* (Collection de travaux de l'Académie internationale d'histoire des sciences, 35). Leiden: Brill.
Newman, William R. 1994a. "The Alchemy of Roger Bacon and the *Tres epistolae* Attributed to Him." In *Comprendre et maîtriser la nature au Moyen Âge, mélanges d'histoire des sciences offerts à Guy Beaujouan*. Geneva: Droz.
Newman, William R. 1994b. "Arabo-Latin Forgeries: The Case of the *Summa perfectionis* (With the Text of Jābir Ibn Ḥayyān's *Liber Regni*)." In Gül A. Russell (ed.), *The "Arabick" Interest of the Natural Philosophers in Seventeenth-Century England*. Leiden: Brill.
Newman, William R. 1995. "The Philosophers' Egg: Theory and Practice in the Alchemy of Roger Bacon." In *Micrologus 3, Le crisi dell'alchimia/The Crisis of Alchemy*. Florence: SISMEL – Edizioni del Galluzzo.
Newman, William R. 1997. "An overview of Roger Bacon's alchemy." In Jeremiah Hackett (ed.), *Roger Bacon and the Sciences. Commemorative Essays*. Leiden: Brill.
Newman, William R. 2000. "Alchemy, Assaying, and Experiment." In Frederic L. Holmes and Trevor H. Levere (eds), *Instruments and Experimentation in the History of Chemistry*. Cambridge, MA: MIT Press.
Newman, William R. 2004. *Promethean Ambitions. Alchemy and the Quest to Perfect Nature*. Chicago, IL: University of Chicago Press.
Newman, William R. 2006. *Atoms and Alchemy: Chymistry and the Experimental Origins of the Scientific Revolution*. Chicago, IL: University of Chicago Press.
Newman, William R. 2013. "Medieval Alchemy." In David C. Lindberg and Michael H. Shank (eds), *The Cambridge History of Science. Vol. 2. Medieval Science*. Cambridge: Cambridge University Press.
Newman, William R. 2014. "Mercury and Sulphur among the High Medieval Alchemists: From Rāzī and Avicenna to Albertus Magnus and Pseudo-Roger Bacon." *Ambix*, 61: 327–44.
Newman, William R., and Lawrence M. Principe. 1998. "Alchemy vs. Chemistry: The Etymological Origins of a Historiographic Mistake." *Early Science and Medicine*, 3: 32–65.
Newman, William R., and Lawrence M. Principe. 2002. *Alchemy Tried in the Fire: Starkey, Boyle, and the Fate of Helmontian Chymistry*. Chicago, IL: University of Chicago Press.
Norris, John A. 2006. "The Mineral Exhalation Theory of Metallogenesis in Pre-Modern Mineral Science." *Ambix*, 53: 43–65.
North, John. 2004. "Diagram and Thought in Medieval Science." In Marie-Thérèse Zenner (ed.), *Villard's Legacy: Studies in Medieval Technology, Science, and Art in Memory of Jean Gimpel*. Aldershot: Ashgate.
Nummedal, Tara E. 2007. *Alchemy and Authority in the Holy Roman Empire*. Chicago, IL: University of Chicago Press.
Obrist, Barbara. 1982. *Les débuts de l'imagerie alchimique (XIVe–XVe siècles)*. Paris: Le Sycomore.
Obrist, Barbara. 1990. *Constantine of Pisa, The Book of the Secrets of Alchemy: Introduction, Critical Edition, Translation and Commentary* (Collection des travaux de l'Académie internationale d'histoire des sciences, 34). Leiden: Brill.
Obrist, Barbara. 1993. "Cosmology and Alchemy in an Illustrated 13th-Century Alchemical Tract: Constantine of Pisa, 'The Book of the Secrets of Alchemy.'" *Micrologus*, 1: 115–60.

Obrist, Barbara. 1996. "Art et nature dans l'alchimie médiévale." *Revue d'histoire des sciences*, 49: 215–86.

Obrist, Barbara. 2003. "Visualisation in Medieval Alchemy." *Hyle: International Journal of the Philosophy of Chemistry*, 9: 131–70.

Obrist, Barbara. 2005. "Alchemy and Secret in the Latin Middle Ages." In Dominique de courcelles (ed.), *D'un principe philosophique à un genre littéraire: les « secrets »*. Paris: Champion.

Ogrinc, Will H.L. 1980. "Western Society and Alchemy from 1200–1500." *Journal of Medieval History*, 6: 103–37.

Osten, Sigrid von. 1998. *Das Alchemistenlaboratorium Oberstockstall. Ein Fundkomplex des 16. Jahrhunderts aus Niederösterreich* (Monographien zur Frühgeschichte und Mittelalterarchäologie, 6). Innsbruck: Universitätsverlag Wagner.

Page, William, and John Horace Round. 1907. *The Victoria History of the County of Essex*, vol. II. London: Constable.

Papathanassiou, Maria. 2005. "L'œuvre alchimique de Stéphanos d'Alexandrie: structure et transformations de la matière, unité et pluralité, l'énigme des philosophes." In Cristina Viano (ed.), *L'alchimie et ses racines philosophiques: la tradition grecque et la tradition arabe*. Paris: Vrin.

Papathanassiou, Maria. 2006. "Stephanos of Alexandria: A Famous Byzantine Scholar, Alchemist and Astrologer." In Paul Magdalino and Maria Mavroudi (eds), *The Occult Sciences in Byzantium*. Geneva: La Pomme d'Or.

Papathanassiou, Maria. 2017. *Stephanos von Alexandreia und sein alchemistisches Werk. Die kritische Edition des griechischen Textes eingeschlossen*. Athens: Cosmoware.

Pappacena, Massimo. 2000. "Teologia e cosmologia in un trattato ermetico in lingua araba: il *Kitāb sirr al-khalīqa*." Ph.D. thesis, Università degli Studi di Napoli L'Orientale.

Paravicini Bagliani, Agostino. 2003. "Ruggero Bacone e l'alchimia di lunga vita. Riflessioni sui testi." In Chiara Crisciani and Agostino Paravicini Bagliani (eds), *Alchimia e medicina nel Medioevo*. Florence: SISMEL – Edizioni del Galluzzo.

Patai, Raphael. 1994. *The Jewish Alchemists. A History and Source Book*. Princeton, NJ: Princeton University Press.

Patar, Benoît (ed.). 1991. "*Iohannes Buridanus, Quaestiones in Aristotelis de anima*." In *Le traité de l'âme de Jean Buridan* (Philosophes médiévales, 29). Louvain-la-Neuve: Longueil (Québec): Éditions du Préambule.

Patar, Benoît (ed.). 1995. *Nicolaus Oresme, Expositio et Quaestiones in Aristotelis De anima*. Louvain-la-Neuve: Éditions Peeters.

Percy, John. 1870. *Metallurgy, Vol. III: Lead, Including Extraction of Silver from Lead*. London: Murray.

Pereira, Michela. 1989. *The Alchemical Corpus Attributed to Raymond Lull* (Warburg Institute Surveys and Texts, 18). London: Warburg Institute, University of London.

Pereira, Michela. 1992. *L'oro dei filosofi: saggio sulle idee di un alchimista del Trecento* (Biblioteca di "Medioevo latino," 7). Spoleto: Centro italiano di studi sull'alto Medioevo.

Pereira, Michela. 1995a. "Le figure alchemiche pseudolulliane: un indice oltre il testo?" In Claudio Leonardi, Marcello Morelli, and Francesco Santi (eds), *Fabula in tabula. Una storia degli indici dal manoscritto al testo elettronico*. Spoleto: Centro italiano di studi sull'alto Medioevo.

Pereira, Michela. 1995b. "Teorie dell'elixir nell'alchimia latina medievale." In *Micrologus 3, Le crisi dell'alchimia/The Crisis of Alchemy*. Florence: SISMEL – Edizioni del Galluzzo.

Pereira, Michela. 1998. "*Mater Medicinarum*: English Physicians and the Alchemical Elixir in the Fifteenth Century." In Roger French, Jon Arrizabalaga, Andrew Cunningham, and Luis Garcia-Ballester (eds), *Medicine from the Black Death to the French Disease*. Aldershot: Ashgate.

Pereira, Michela. 1999. "Alchemy and the Use of Vernacular Languages in the Late Middle Ages." *Speculum*, 74: 336–56.

Pereira, Michela. 2010. *Arcana Sapientia: L'Alchemia dalle origini a Jung*. Rome: Carocci.

Pereira, Michela, and Barbara Spaggiari. 1999. *Il Testamentum alchemico attribuito a Raimondo Lullo. Edizione del testo latino e catalano dal manoscritto Oxford, Corpus Christi College, 244* (Millennio medievale, 14, Testi 6). Florence: SISMEL – Edizioni del Galluzzo.

Pertsch, Wilhelm. 1880. *Die orientalischen Handschriften der Herzoglichen Bibliothek zu Gotha. Theil 3: Die arabischen Handschriften,* Vol. II. Gotha: Perthes.

Phelps Matthew, Ian Charles Freestone, Yael Gorin-Rosen, and Bernard Gratuze. 2016. "Natron Glass Production and Supply in the Late Antique and Early Medieval Near East: The Effect of the Byzantine-Islamic Transition." *Journal of Archaeological Science*, 75: 57–71.

Pingree, David Edwin (ed.). 1986. Maslama ibn Qāsim al-Qurṭubī (Pseudo-Majrīṭī). *Picatrix, The Latin Version of the Ghāyat al-Ḥakīm* (Studies of the Warburg Institute, 39). London: Warburg Institute.

Plessner, Martin. 1975. *Vorsokratische Philosophie und griechische Alchemie in arabisch-lateinischer Überlieferung. Studien zu Text und Inhalt der* Turba philosophorum (Boethius: Texte und Abhandlungen zur Geschichte der exakten Wissenschaften, 4). Wiesbaden: Franz Steiner.

Pormann, Peter E., and Emilie Savage-Smith. 2007. *Medieval Islamic Medicine*. Washington, DC: Georgetown University Press.

Principe, Lawrence M. 1987. "'Chemical Translation' and the Role of Impurities in Alchemy: Examples from Basil Valentine's Triumph-Wagen." *Ambix*, 34: 21–30.

Principe, Lawrence M. 1998. *The Aspiring Adept: Robert Boyle and His Alchemical Quest*. Princeton, NJ: Princeton University Press.

Principe, Lawrence M. 2013. *The Secrets of Alchemy*. Chicago, IL: University of Chicago Press.

Principe, Lawrence M. 2016. "Chymical Exotica in the Seventeenth Century, or, How to Make the Bologna Stone." *Ambix*, 63: 118–44.

Principe, Lawrence M., and William R. Newman. 2001. "Some Problems with the Historiography of Alchemy." In William R. Newman and Anthony Grafton (eds), *Secrets of Nature. Astrology and Alchemy in Early Modern Europe*. Cambridge, MA: MIT Press.

Pseudo-Aristotle. 1555. *Secretum secretorum Aristotelis ad Alexandrum Magnum*, ed. Francesco Storella, transl. Philip of Tripoli. Venice.

Ramage, Andrew, and Paul T. Craddock. 2000. *King Croesus' Gold: Excavations at Sardis and the History of Gold Refining* (Archaeological Exploration of Sardis, 11). London: British Museum Press.

Rampling, Jennifer M. 2010. "The Catalogue of the Ripley Corpus: Alchemical Writings Attributed to George Ripley (d. ca. 1490)." *Ambix*, 57: 125–201.

Rampling, Jennifer M. 2012. "Transmission and Transmutation: George Ripley and the Place of English Alchemy in Early Modern Europe." *Early Science and Medicine*, 17: 477–99.
Rampling, Jennifer M. 2013. "Depicting the Medieval Alchemical Cosmos: George Ripley's *Wheel of Inferior Astronomy*." *Early Science and Medicine*, 18: 45–86.
Rampling, Jennifer M. 2014a. "A Secret Language: The Ripley Scrolls." In Sven Dupré, Dedo von Kerssenbrock-Krosigk, and Beat Wismer (eds), *Art and Alchemy. The Mystery of Transformation*. Munich: Hirmer.
Rampling, Jennifer M. 2014b. "Transmuting Sericon: Alchemy as 'Practical Exegesis' in Early Modern England." *Osiris*, 29: 19–34.
Rampling, Jennifer M. 2018. "How to Sublime Mercury: Reading Like a Philosopher in Medieval Europe." "History of Knowledge" (blog). Available online: https://historyofknowledge.net/2018/05/24/reading-like-a-philosopher/ (accessed August 25, 2021).
Rampling, Jennifer M. 2020. *The Experimental Fire. Inventing English Alchemy*. Chicago, IL: University of Chicago Press.
Ray, Praphulla Chandra. 1956. *History of Chemistry in Ancient and Medieval India*. Calcutta: Indian Chemical Society.
Reeves, Marjorie, and Beatrice Hirsch-Reich. 1972. *The Figurae of Joachim of Fiore*. Oxford: Clarendon Press.
Rehren, Thilo, and Sam Nixon. 2014. "Refining Gold with Glass – an Early Islamic Technology at Tadmekka, Mali." *Journal of Archaeological Science*, 49: 33–41.
Rehren, Thilo, Sam Nixon, and Maria Filomena Guerra. 2011. "New Light on the Early Islamic West African Gold Trade: Coin Moulds from Tadmekka, Mali." *Antiquity*, 85(330): 1353–68.
Reichert, Benedictus Maria. 1896. *Fratris Gerardi de Fracheto O. P. Vitae Fratrum Ordinis Praedicatorum necnon Chronica Ordinis ab anno MCCIII usque ad MCCLIV*. Rome: Istituto storica domenicano.
Rigius, Ludovicus. 1488. *Opusculum praeclarum beati Thomae Aquinatis quod de esse et essentiis tum realibus tum intentionalibus inscribitur*. Venice.
Rodríguez Guerrero, José. 2010. "Some Forgotten Fez Alchemists and the Loss of the Peñon de Vélez de la Gomera in the Sixteenth Century." In Miguel López-Pérez, Didier Kahn, and Mar Rey-Bueno. (ed.), *Chymia: Science and Nature in Medieval and Early Modern Europe*. Newcastle-upon-Tyne: Cambridge Scholars Publishing.
Rodríguez Guerrero, José. 2010–2013. "Un Repaso a la Alquimia del Midi Francés en el Siglo XIV (Parte I)." *Azogue, Revista Electrónica dedicada al Estudio Histórico-Crítico de la Alquimia*, 7: 75–141. Available online: http://www.revistaazogue.com/Azogue7-3.pdf (accessed August 25, 2021).
Rosenthal, Franz. 1947. *The Technique and Approach of Muslim Scholarship*. Rome: Pontificium Institutum Biblicum.
Rouaze, Isabelle. 1986. "Un atelier de distillation au Moyen Âge." *Bulletin archéologique du Comité des travaux historiques et scientifiques, Nouvelle série*, 22: 159–271.
Rubino, Elisa, and Samuela Pagani. 2016. "Il *De mineralibus* di Avicenna tradotto da Alfredo di Shareshill." *Bulletin de philosophie médiévale*, 58: 23–87.
Rudolph, Ulrich. 1990. "Christliche Theologie und vorsokratische Lehren in der 'Turba Philosophorum.'" *Oriens*, 32: 97–123.
Rudolph, Ulrich. 2012a. "Einleitung." In Ulrich Rudolph (ed.), *Philosophie in der islamischen Welt. Vol. I: 8–10. Jahrhundert*. Basel: Schwabe.

Rudolph, Ulrich. 2012b. "Abū Naṣr al-Fārābī." In Ulrich Rudolph (ed.), *Philosophie in der islamischen Welt. Vol. I: 8–10. Jahrhundert*. Basel: Schwabe.

Ruska, Julius. 1923. "Chemische Apparatur bei den Arabern und Persern und im Abendland am Ausgang des Mittelalters." *Chemische Apparatur*, 10: 137–9.

Ruska, Julius. 1924. *Arabische Alchemisten. I. Chālid ibn Jazīd ibn Mu'āwija* (Heidelberger Akten der von-Portheim-Stiftung, 6, Arbeiten aus dem Institut für Geschichte der Naturwissenschaft, I). Heidelberg: Carl Winter's Universitätsbuchhandlung.

Ruska, Julius. 1926. *Tabula Smaragdina. Ein Beitrag zur Geschichte der hermetischen Literatur* (Heidelberger Akten der von-Portheim-Stiftung, 14, Arbeiten aus dem Institut für Geschichte der Naturwissenschaft, IV). Heidelberg: Carl Winter's Universitätsbuchhandlung.

Ruska, Julius. 1928. "Zwei Bücher *De Compositione Alchemiae* und ihre Vorreden." *Archiv für Geschichte der Mathematik, der Naturwissenschaften und der Technik*, 11: 28–37.

Ruska, Julius. 1931. *Turba philosophorum: ein Beitrag zur Geschichte der Alchemie* (Quellen und Studien zur Geschichte der Naturwissenschaften und der Medizin, 1). Berlin: Springer.

Ruska, Julius. 1935. "Übersetzung und Bearbeitungen von al-Rāzīs Buch *Geheimnis der Geheimnisse*." *Quellen und Studien zur Geschichte der Naturwissenschaften und der Medizin*, 4(3): 153–238.

Ruska, Julius. 1936. "Studien zu den chemisch-technischen Rezeptsammlungen des *Liber Sacerdotum*." *Quellen und Studien zur Geschichte der Naturwissenschaften und der Medizin*, 5: 83–125.

Ruska, Julius. 1937. "Al-Rāzī's Buch *Geheimnis der Geheimnisse*, mit Einleitung und Erlauterungen in deutscher Übersetzung." *Quellen und Studien zur Geschichte der Naturwissenschaften und der Medizin*, 6: 1–246.

Ryan, Michael. 2011. *A Kingdom of Stargazers: Astrology and Authority in the Late Medieval Crown of Aragon*. Ithaca, NY: Cornell University Press.

Sabra, A.I. 1976. "al-Kh̲wārazmī." In *The Encyclopaedia of Islam*. New Edition, vol. IV. Leiden: Brill.

Saif, Liana. 2016. "The Cows and the Bees: Arabic Sources and Parallels for Pseudo-Plato's *Liber Vaccae* (*Kitāb al-Nawāmīs*)." *Journal of the Warburg and Courtauld Institutes*, 79: 1–47.

Sarton, George. 1947. *Introduction to the History of Science IV*. Baltimore, MD: Williams and Wilkins.

Saussus, Lise, and Nicolas Thomas. 2019. *Un atelier d'orfèvre autour de l'An Mil. Travail du cuivre, de l'argent et du fer à Oostvleteren (Flandre-Occidentale, Belgique)* (Collection d'archéologie Joseph Mertens, 18). Louvain-la-Neuve: Presses universitaires de Louvain.

Savage-Smith, Emilie. 1997. "Glass Alchemical Equipment." In Francis Maddison and Emilie Savage-Smith (eds), *Science, Tools and Magic*. Part One: Body and Spirit, Mapping the Universe. London: Nour Foundation in association with Azimuth Editions and Oxford University Press.

Sayili, Aydin. 1951. "Fârâbî'nin Simyanın Lüzumu Hakkındaki Risalesi." *Belleten, Türk Tarih Kurumu*, 15(57): 65–79.

Sayyid, Ayman Fu'ād (ed.). 2009. Ibn al-Nadīm. *Kitāb al-Fihrist. The Fihrist of al-Nadīm*. London: Al-Furqān Islamic Heritage Foundation.

Scafi, Alessandro. 1993. "*Aurum hungaricum*: il re Mathia della Ungheria e il segreto della alchimia." *Rivista di studi ungheresi*, 8: 5–16.

Scarborough, John. 1984. "Early Byzantine Pharmacology." *Dumbarton Oaks Papers*, 38: 213–32.

Schibille, Nadine, Andrew Meek, Bendeguz Tobias, Chris Entwistle, Mathilde Avisseau-Broustet, Henrique Da Mota, and Bernard Gratuze. 2016. "Comprehensive Chemical Characterisation of Byzantine Glass Weights." *PLoS ONE*. Available online: doi.org/10.1371/journal.pone.0168289 (accessed August 25, 2021).

Schibille, Nadine, Andrew Meek, Mark T. Wypyski, Jens Kröger, Mariam Rosser-Owen, and Rosalind Wade Haddon. 2018. "The Glass Walls of Samarra (Iraq): Ninth Century Abbasid Glass Production and Imports." *PLoS ONE*. Available online: doi.org/10.1371/journal.pone.0201749 (accessed August 25, 2021).

Schoeler, Gregor. 2013. "Gesprochenes Wort und Schrift. Mündlichkeit und Schriftlichkeit im frühislamischen Lehrbetrieb." In Peter Gemeinhardt and Sebastian Günther (eds), *Von Rom nach Bagdad. Bildung und Religion von der römischen Kaiserzeit bis zum klassischen Islam*. Tübingen: Mohr Siebeck.

Scholem, Gershom. 1925. *Alchemie und Kabbala*. Breslau: Schatzky.

Scholem, Gershom. 1926. *Sefer ha-Tamar. Das Buch von der Palme des Abū Aflah aus Syracus. Ein Text aus der arabischen Geheimwissenschaft. Nach der allein erhaltenen hebräischen Übersetzung*. Hanover: Lafaire.

Sellner, Christiane, Heribert J. Oel, and Boubacar Camara. 1979. "Untersuchung alter Gläser (Waldglas) auf Zusammenhang von Zusammensetzung, Farbe and Schmelzatmosheäre mit die Elektronenspektroskopie und der Electronenspinresonanz (ESR)." *Glastechnische Berichte*, 52: 255–64.

Sezgin, Fuat. 1971. *Geschichte des arabischen Schrifttums. Band IV (Alchimie – Chemie – Botanik – Agrikultur, bis ca. 430 H.)*. Leiden: Brill.

Sezgin, Fuat, and Eckhard Neubauer. 2010. *Science and Technology in Islam. Vol. 4. Catalogue of the Collection of Instruments of the Institute for the History of Arabic and Islamic Sciences*. Frankfurt: Institut für Geschichte der Arabisch-Islamischen Wissenschaften an der Johann Wolfgang Goethe-Universität.

Shaddoud, Ibrahim. 2017. "Vaisselier de santé dans le monde arabe (VIIIe–XVe siècles): une restitution possible des usages grâce au croisement des sources." In Sergei Bocharov, Véronique François, and Ayrat Sitdikov (eds), *Glazed Pottery of the Mediterranean and the Black Sea Region, 10th–18th Centuries*. Kazan: Stratum Publishing House.

Shapin, Steven. 1988. "The House of Experiment in Seventeenth-Century England." *Isis*, 79: 373–404.

Siemianowska, Sylwia, Aleksandra Pankiewicz, and Krysztof Sadowski. 2019. "On Technology and Production Techniques of Early Medieval Glass Rings from Silesia." *Archaeometry*, 61: 614–46.

Siggel, Alfred. 1949–1956. *Katalog der arabischen alchemistischen Handschriften Deutschlands*. 3 vols. Berlin: Akademie-Verlag.

Siggel, Alfred. 1951. *Decknamen in der arabischen alchemistischen Literatur*. Berlin: Akademie-Verlag.

Sisco, Anneliese Grünhaldt, and Cyril Stanley Smith (ed. and tr.). 1951. *Lazarus Ercker's Treatise on Ores and Assaying, Translated from the German Edition of 1580*. Chicago: University of Chicago Press.

Sivin, Nathan. 1968. *Chinese Alchemy. Preliminary Studies*. Cambridge, MA: Harvard University Press.

Sivin, Nathan. 1990. "Research on the History of Chinese Alchemy." In Zweder R.W.M. von Martels (ed.), *Alchemy Revisited. Proceedings of the International Conference on the History of Alchemy at the University of Groningen, 17–19 April 1989*. Leiden: Brill.

Smith, Cyril Stanley, and John G. Hawthorne. 1974. "*Mappae clavicula*. A Little Key to the World of Medieval Techniques." *Transactions of the American Philosophical Society*, 64(4): 1–128.

Smith, Cyril Stanley, and Martha Teach Gnudi. 1966. *The Pirotechnia of Vannoccio Biringuccio*. Cambridge, MA: MIT Press.

Soukup, Rudolf Werner, and Helmut Mayer. 1997. *Alchemistisches Gold, paracelsistische Pharmaka. Laboratoriumstechnik im 16. Jahrhundert; Chemiegeschichtliche und archäometrische Untersuchungen am Inventar des Laboratoriums von Oberstockstall/Kirchberg am Wagram* (Perspektiven der Wissenschaftsgeschichte, 10). Vienna: Böhlau.

Stapleton, Henry Ernest, and Rizkallah F. Azo. 1905. "Alchemical Equipment in the Eleventh Century A.D." *Memoirs of the Asiatic Society of Bengal*, 1(4): 47–70.

Stapleton, Henry Ernest, M. Hidāyat Ḥusain, and M. Turāb 'Alī. 1933. "Three Arabic Treatises on Alchemy by Muḥammad Bin Umayl (10th Century A.D.)." *Memoirs of the Asiatic Society of Bengal*, 12(1): 1–213.

Stapleton, Henry Ernest, Rizkallah F. Azo, and M. Hidāyat Ḥusain. 1927. "Chemistry in 'Irāq and Persia in the Tenth Century A. D." *Memoirs of the Asiatic Society of Bengal*, 8(6): 317–418.

Stavenhagen, Lee (ed.). 1974. *A Testament of Alchemy, being the Revelations of Morienus, Ancient Adept and Hermit of Jerusalem to Khalid ibn Yazid ibn Mu'awiyya, King of the Arabs of the Divine Secrets of the Magisterium and Accomplishment of the Alchemical Art*. Hanover, NH: University Press of New England for the Brandeis University Press.

Steele, Robert, and Dorothea Waley Singer. 1928. "The Emerald Tablet." *Proceedings of the Royal Society of Medicine*, 21: 485–501.

Steinschneider, Moritz. 1873. "Typen." *Jeshurun. Zeitschrift für die Wissenschaft des Judenthums*, 9: 84.

Storey, Charles Ambrose. 1977. *Persian Literature. A Bio-bibliographical Survey. Vol. II, part 3: F. Encyclopedias and Miscellanies. G. Arts and Crafts. H. Science. J. Occult Arts*. London: Royal Asiatic Society of Great Britain and Ireland.

Strohmaier, Gotthard. 1991. "'Umāra ibn Ḥamza, Constantine V, and the Invention of the Elixir." In Vassilios Christides et al. (eds), *Second and Third International Congress on Greek and Arabic Studies*. Athens: Ministry of Culture.

Strohmaier, Gotthard. 2016. "Elixir, Alchemy and the Metamorphoses of Two Synonyms." *Al-Qanṭara*, 37: 423–34.

Stüve, Wilhelm. 1900. *Olympiodori In Aristotelis Meteora Commentaria*. Berlin: G. Reimer.

Szymański, Michał. 2015. "Przyczynek do identyfikacji średniowiecznej aparatury destylacyjnej w Polsce." *Acta Universitatis Lodziensis. Folia Archaeologica*, 30: 207–23.

Tait, Hugh. 1979. *The Golden Age of Venetian Glass*. London: British Museum.

Tait, Hugh. 1999. "Venice: Heir to the Glassmakers of Islam or of Byzantium." In *Islam and the Italian Renaissance* (Warburg Institute Colloquia, 5). London: Warburg Institute.

Tajaddud, Riḍā. 1393 AH/1973 AD. Ibn al-Nadīm. *Kitāb al-Fihrist*. Tehran: Marvi.

Takahashi, Hidemi. 2015. "Syriac as the Intermediary in Scientific Graeco-Arabica: Some Historical and Philological Observations." *Intellectual History of the Islamicate World*, 3: 66–97.

Ṭāshköprüzāde. n.d. *Miftāḥ al-sa'āda wa-miṣbāḥ al-siyāda*, vol. 1. Beirut: Dār al-kutub al-'ilmiyya.

Taylor, Frank Sherwood. 1937. "Alchemical Works of Stephanus of Alexandria, Part I." *Ambix*, 1: 116–39.

Taylor, Frank Sherwood. 1938. "Alchemical Works of Stephanus of Alexandria, Part II." *Ambix*, 2: 39–49.

Taylor, Frank Sherwood. 1945. "The Evolution of the Still." *Annals of Science*, 5: 185–202.

Taylor, Frank Sherwood, and Charles Singer. 1956. "Pre-scientific Industrial Chemistry." In Charles Singer, Eric John Holmyard, A. Rupert Hall, and Trevor I. Williams (eds), *A History of Technology. Vol. 2: The Mediterranean Civilizations and the Middle Ages, c. 700* B.C. *to c.* A.D. *1500*. Oxford: Clarendon Press.

Telle, Joachim. 1980. *Sol und Luna. Literatur- und alchemiegeschichtliche Studien zu einem altdeutschen Bildgedicht*. Hürtgenwald: Pressler.

Telle, Joachim. 1992. *Rosarium philosophorum. Ein alchemistisches Florilegium des Spätmittelalters. Faksimile der illustrierten Erstausgabe*. 2 vols. Weinheim: VCH.

Theissen, Wilfred R. 1986. "John Dastin's Letter on the Philosophers' Stone." *Ambix*, 33: 78–87.

Theissen, Wilfred R. 1991. "John Dastin: The Alchemist as Co-Creator." *Ambix*, 38: 73–8.

Theissen, Wilfred R. 1999. "John Dastin's Alchemical Vision." *Ambix*, 46: 65–72.

Theissen, Wilfred R. 2008. "The Letters of John Dastin." *Ambix*, 55: 153–68.

Thiriot, Jacques. 1997. "Les fours pour la préparation des glaçures dans le monde méditerranéen." In Gabrielle Démians d'Archimbaud (ed.), *La Céramique Médiévale en Méditerranée. Actes du VI*[e] *congrès de l'AIECM2. Aix-en-Provence, 13–18 novembre 1995*. Aix-en-Provence: Narration.

Thomas, Nicolas. 2006. "Prendre de l'acier pour de l'or." In *Hypothèses 2005*. Paris: Publications de la Sorbonne.

Thomas, Nicolas. 2009. "L'alambic dans la cuisine?" In Fabienne Ravoire and Anne Dietrich (eds), *La cuisine et la table dans la France de la fin du Moyen Âge. Contenus et contenants du XIV*[e] *au XVI*[e] *siècle. Actes du colloque de Sens (2004)*. Caen: Publications du CRAHM.

Thomas, Nicolas. 2013. "De la recette à la pratique, l'exemple du *lutum sapientiae* des alchimistes." In Ricardo Córdoba De La Llave (ed.), *Craft Treatises and Handbooks. The Dissemination of Technical Knowledge in the Middle Ages – International Symposium Córdoba*. Turnhout: Brepols.

Thomas, Nicolas. 2020. "*Aqua vitae et aqua ardens*. Production et consommation des produits distillés de boissons fermentées (XIIe–XVIIe siècle)." *Archéopages*, 47: 58–63.

Thomas, Nicolas, and Caroline Claude. 2011. "Les vases à fond percé: Pratique de la distillation *per descensum* au bas Moyen Âge en Île-de-France." *Revue archéologique d'Île-de-France*, 4: 267–88.

Thomas, Nicolas, and Nicole Rodrigues. 2017. "Le verre dans la médecine et dans le laboratoire alchimique." In Sophie Lagabrielle (ed.), *Le verre: Un Moyen Âge inventif [Catalogue de l'exposition tenu au musée de Cluny – musée national du Moyen Âge, à Paris, du 20 septembre 2017 au 8 janvier 2018]*. Paris: Réunion des musées nationaux.

Thomas, Phillip D. 1968. *David Ragor's Transcription of Walter of Odington's "Icocedron"* (Wichita State University Bulletin, 44.3). Wichita, KS: Wichita State University.

Thomas, Phillip Drennon. 1971. "The Alchemical Thought of Walter of Odington." In *Actes du XIIe congrès international d'histoire des sciences. Tome 3A. Science et philosophie: Antiquité, Moyen Âge, Renaissance*. Paris: Blanchard.
Thompson, Daniel V. 1960. *Cennino Cennini's The Craftsman's Handbook*. New York: Dover.
Thomson, S. Harrison. 1938. "The Texts of Michael Scot's *Ars Alchemie*." *Osiris*, 5: 523–59.
Thorndike, Lynn. 1923–1958. *A History of Magic and Experimental Science*. New York: Columbia University Press.
Timmermann, Anke. 2013. *Verse and Transmutation: A Corpus of Middle English Alchemical Poetry*. Leiden: Brill.
Todd, Richard. 2016. "Alchemical Poetry in Almohad Morocco: The *Shudhūr al-dhahab* of Ibn Arfaʿ Ra's." *Oriens*, 44: 116–44.
Toll, Christopher (ed.). 1968. al-Hamdānī, al-Ḥasan ibn Aḥmad. *Al-Jawharatayn al-ʿatīqatayn al-māʾiʿatayn al-ṣafrāʾ wa-l-bayḍāʾ. Die beiden Edelmetalle Gold und Silber* (Acta Universitatis Upsaliensis. Studia Semitica Upsaliensia, 1). Uppsala: Almqvist & Wiksell.
Travaglia, Pinella. 2001. *Una cosmologia ermetica, Il Kitāb sirr al-ḫalīqa/De secretis naturae*. Naples: Liguori.
Tylecote, Ronald Frank. 1986. *The Prehistory of Metallurgy in the British Isles*. London: Institute of Metals.
Tyson, Rachel Caroline. 1996. "Medieval Glass Vessels in England AD 1200–1500: A Survey." Ph.D. thesis, Durham University. Available online: http://etheses.dur.ac.uk/1223/.
Tyson, Rachel Caroline. 2000. *Medieval Glass Vessels Found in England c AD 1200–1500* (Research Report 121). York: Council for British Archaeology.
Ullmann, Manfred. 1972. *Die Natur- und Geheimwissenschaften im Islam* (Handbuch der Orientalistik, erste Abteilung, Ergänzungsband VI zweiter Abschnitt). Leiden: Brill.
Ullmann, Manfred. 1974–1976. *Katalog der arabischen alchemistischen Handschriften der Chester Beatty Library*. 2 vols. Wiesbaden: Harrassowitz.
Ullmann, Manfred. 1978. "Ḫālid ibn Yazīd und die Alchemie: Eine Legende." *Der Islam, Zeitschrift für Geschichte und Kultur des islamischen Orients*, 55: 181–218.
Ullmann, Manfred. 1986. "Kīmiyāʾ." In *Encyclopedia of Islam, New Edition*. Leiden: Brill.
Valiulina, Svetlana. 2016. "Medieval Workshop of an Alchemist, Jeweller and Glassmaker in Bilyar (Middle Volga Region, Russian Federation)." *Památky Archeologické*, 107: 237–78.
Vallauri, Lucy, and Marie Leenhardt. 1997. "Les productions céramiques." In Henri Marchesi, Jacques Thiriot, and Lucy Vallauri (eds), *Marseille, les ateliers de potiers du XIIIe s. et le quartier Sainte-Barbe (V^e^–XVIIe s.)* (Document d'archéologie française; 65). Paris: Éditions de la Maison des sciences de l'homme.
Vereno, Ingolf. 1992. *Studien zum ältesten alchemistischen Schrifttum: auf der Grundlage zweier erstmals edierter arabischer Hermetica* (Islamkundliche Untersuchungen, 155). Berlin: Klaus Schwarz.
Viano, Cristina. 1995. "Olympiodore l'alchimiste et les Présocratiques. Une doxographie de l'unité (*De arte sacra*, §§ 18–27)." In Didier Kahn and Sylvain Matton (eds), *Alchymie: art, histoire et mythes*. Paris: S.É.H.A.-ARCHÈ.
Viano, Cristina. 1996. "Aristote et l'alchimie grecque: la transmutation et le modèle aristotélicien entre théorie et pratique." *Revue d'histoire des sciences*, 49: 189–213.

Viano, Cristina. 2006. *La matière des choses. Livre IV des météorologiques d'Aristote et son interprétation par Olympiodore*. Paris: Vrin.
Vincent of Beauvais. 1624. *Bibliotheca mundi: Speculum quadruplex, naturale, doctrinale, morale, historiale*. Douai: Bellerius.
Vloten, Gerlof van (ed.). 1968. Khwārizmī. *Liber Mafātīḥ al-'ulūm: explicans vocabula technica scientiarum tam Arabum quam peregrinorum*, 2nd ed. Leiden: Brill.
Völlnagel, Jörg. 2004. *Splendor solis oder Sonnenglanz: Studien zu einer alchemistischen Bilderhandschrift*. Munich: Deutscher Kunstverlag.
Völlnagel, Jörg. 2011. *Harley MS. 3469: Splendor solis or Splendour of the Sun: A German Alchemical Manuscript*. London: British Library.
Voskoboynikov, Oleg (ed.). 2019. *Michel Scot, Liber particularis et Liber physonomie* (Micrologus' Library, 93). Florence: SISMEL – Edizioni del Galluzzo.
Weisser, Ursula. 1979. *Buch über das Geheimnis der Schöpfung und die Darstellung der Natur (Buch der Ursachen) von Pseudo-Apollonios von Tyana* (Sources and Studies in the History of Arabic-Islamic Science, 1). Aleppo: Institute for the History of Arabic Science, University of Aleppo.
Weisser, Ursula. 1980. *Das "Buch über das Geheimnis der Schöpfung" von Pseudo-Apollonios von Tyana* (Ars medica: Texte und Untersuchungen zur Quellenkunde der alten Medizin, 2). Berlin: De Gruyter.
Wiedemann, Eilhard. 1909. "Über chemische Apparate bei der Arabern." In Paul Diergart (ed.), *Beiträge aus der Geschichte der Chemie*. Leipzig: Deuticke.
Wild, Mike, and Ian Eastwood. 1992. "Soil Contamination and Smelting Sites." In Lynn Willies and David Cranstone (eds), *Boles and Smelt Mills*. London: Historical Metallurgy Society.
Williams, Alan. 2007. "Crucible Steel in Medieval Swords." In Susan La Niece, Duncan Hook, and Paul T. Craddock (eds), *Metals and Mines: Studies in Archaeometallurgy*. London: Archetype Publications.
Willmott, Hugh. 2002. *Early Post-Medieval Vessel Glass in England, c. 1500–1670* (Council for British Archaeology Research Reports, 132). York: Council for British Archaeology.
Wilson, Malcolm. 2013. *Structure and Method in Aristotle's* Meteorologica. *A More Disorderly Nature*. Cambridge: Cambridge University Press.
Witkam, Januarius Justus. 1989. "De egyptische arts Ibn al-Akfānī (gest. 749/1348) en zijn indeling van de wetenschappen. Editie van het *Kitāb iršād al-qāṣid ilā asnā al-maqāṣid*." Ph.D. thesis, University of Leiden.
Wujastyk, Dominik. 1984. "An Alchemical Ghost: The Rasaratnâkara by Nâgârjuna." *Ambix*, 31: 70–83.
Wyckoff, Dorothy. 1967. *Albertus Magnus. Book of Minerals*. Oxford: Clarendon Press.
Yamamoto, Keiji, and Charles Burnett. 2000. *Abū Ma'shar on Historical Astrology. The Book of Religions and Dynasties (On Great Conjunctions). Vol. 1. The Arabic Original: Abū Ma'shar, Kitāb al-milal wa-d-duwal (The Book of Religions and Dynasties)* (Islamic Philosophy, Theology and Science, 33). Leiden: Brill.
Yinon, Yosef (Fenton, Paul B.). 1993. "R. Makhluf Amsalem, an Alchemist and Kabbalist of Morocco." *Pe'amim*, 55: 92–123 [in Hebrew].
Zetzner, Lazarus. 1602–1661. *Theatrum chemicum*. Strasbourg: Lazare Zetzner et ses héritiers.

LIST OF CONTRIBUTORS

Justine Bayley, University College London

Spike Bucklow, Hamilton Kerr Institute, University of Cambridge

Charles Burnett, Warburg Institute, University of London

Antoine Calvet, independent researcher

Gabriele Ferrario, Università di Bologna

Regula Forster, Eberhard Karls Universität Tübingen

Jean-Marc Mandosio, École Pratique des Hautes Études

Matteo Martelli, Università di Bologna

Sébastien Moureau, FNRS – University of Louvain

Jennifer M. Rampling, Princeton University

Nicolas Thomas, Institut National de Recherches Archéologiques Préventives – Laboratoire de médiévistique occidentale de Paris (University Paris 1 Panthéon-Sorbonne-CNRS, UMR 8589)

INDEX

Note that no distinction is made between genuine and apocryphal (pseudo) authors

'Abd al-Jabbār al-Hamadhānī 136
'Abd al-Raḥmān III, caliph 99
Abū 'Abd Allāh Muḥammad b. Aḥmad al-Kātib al-Khwārizmī 50, 71
Abū Ma'shar 94
Adam, first man 8
Adam of Bodenstein 148
Aëtius of Amida 18
Africanus *see* Leo
Agathodaimon 4, 137
Agricola 112
Akhmīm, Egyptian city (Panopolis) 133
Albert the Great 13, 22, 25, 29, 52, 63, 85, 95, 100–1, 143–4, 147
alchemical operations 40–7
alchemical texts 6–15, 23, 89, 137, 140–6
  attribution of alchemical texts 4, 11–14, 135, 143–4
  hidden language 142
alchemical thinking, emergence of 17
alchemists as individuals 95
alchemy
  alchemy and other branches of knowledge 152
  defenders of 77, 80, 101
  definition of 1–3
  difficulties with the study of 3–5, 36, 142
  distinguishing features of 141
  epistemological status 175
  main goal of 35–6
  moral dangers 96–8
  opponents of 77, 80, 101
  origins of 6, 162
  practical or *speculative* 84
  in relation to other scholarship 71
  teaching and learning of 131, 134, 138–9, 145–6
  theoretical 36
  true and false 86; *see also* Arabic alchemy
alcohol 15, 40–41, 47, 52, 86, 109–10
Alderotti, Taddeo 65, 86
alembic 40, 52–7, 62, 64–6, 68, 171
Alexander of Hales 105–6
Alfred of Shareshill (Alfred the Englishman) 13, 85–6
'Alī b. Abī Ṭālib 8, 79
'Alī b. Ṭāwūs al-Ḥillī 73
allegory and allegorized procedures 3, 12, 37, 53, 132, 140, 160–8, 171, 173
aludel 45, 50, 52–56, 61–2, 64, 66–7
Anderson, Robert 58–9
d'Andrea, Giovanni 103
Anepigraphos, philosopher 19, 24, 26
animal products 110
Antichrist 98
apocryphal compositions 7
Apollonius of Tyana 4, 20, 22, 73, 132, 161, 180

apparatus *see* equipment
aqua ardens see alcohol
Aquinas, Thomas 4, 13, 81, 84, 97, 100, 144, 147, 167
Arabic alchemy 3, 6–10, 13–14, 20–1, 31, 75, 99, 131–4, 139–40, 152–3, 160–4, 168, 173–5
  Greek and Persian influence on 7, 24, 27
  transmission to the Latin West 10–12, 80, 139, 162
Arabic language 67
archaeological sources 37, 55–9, 69
archaeometry 70
Archelaos 7
Arisleus 140
Aristotle and Aristotelian theory 4, 6, 11, 13, 18, 20, 23, 31, 33, 72, 74, 77, 84–5, 97, 101, 135, 140, 142, 146–7, 154–5, 183
Arnald of Villanova (Arnau de Villeneuve) 14, 30, 52, 83–4, 86, 103, 105, 144, 147–8, 165
art
  alchemical 151
  mechanical or *liberal* 81–3
artisanal chemistry 138
Ashmole, Elias 173
astrology, astrological 72–3, 80, 82, 94, 148
  astrological charts 154
Aurora consurgens 164, 167–76
Avicenna *see* Ibn Sīnā
Azo, R.F. 50

Bacon, Roger 2, 14, 29, 83–6, 95, 100, 143, 147, 180
al-Baghdādī, ʻAbd al-Laṭīf 77, 138
Baḥyah ibn Paqudah 91
balance, science of the 21, 28, 152, 154; *see also* chemist's balance
Balīnās/Balīnūs 7, 152, 161, 180; *see also* Apollonius
banking systems 107–8
Bernard de Gordon 86
Bessemer's Converter 128
Bilyar, Russia 55
Biringuccio, Vannocio 112
blast furnaces 114, 128
Blemmydes 18, 26
bloomeries 114–15
body–spirit classification 25
Bonaventura, St 96–7
Bonaventura of Iseo 100
Boniface VIII, pope 86
Bonus, Petrus (Pietro Bono) 15, 84–5, 87, 140, 145, 165
Book of the Fourths 7
Borgognoni, Teodorico 86
botus barbatus: *see* descensory
Brant, Sebastian 98
brass-making 125–6
Brethren of Purity 72
Brueghel, Peter the Elder 63
Buddha statue 129
Bunschwig, Hieronymus 63
Buridan, Johannes 96
Byzantine Empire 18–26, 119

Cairo Genizah 88–9, 91
calcination 38, 43, 167
Carmelite Order 101
Carusi, Paola 74
cast iron 115
cementation 40, 45, 47
ceration 43
chaîne opératoire processes 111
Charles V, king of France 99, 147
Chaucer, Geoffrey 96, 101
chemical and physical changes 110–11
chemical processes 109–12, 128–9, 170
  documentation of 129
  examples of their use 109, 111
  represented by human figures 164
'chemical wedding' image 163–7, 173
chemist's balance 99
chemistry
  medieval 1–2
  in the modern sense 1–2, 16
chimina see descensory
China and Chinese alchemy 4, 75
chimia 1–2
Christ Church, Oxford 58
Christian, Christianity 14, 18, 82, 90, 100, 135, 140, 142, 164–7, 170, 173, 175, 183; *see also* Jesus Christ
  Christian kabbalists 88, 95
Christianus, philosopher 24, 27
Christine de Pizan 147

Christopher of Paris 160
church institutions 100–3
Chymes 4
circulation: *see* cohobation
Cistercian Order 101
Clareno, Angelo 101
Clement IV, pope 95
Clement IX, pope 87
coagulation 39–40, 44
coded language 5, 37, 135, 138, 145; *see also* Decknamen
cohobation 41–2, 60
coinage 99, 126–7
coloring 122–6
Compositiones variae 10
Compound of Alchymie 15
Constantine of Pisa 153–4
Constantine V Copronymos, Byzantine emperor 139
continuity of craft processes 128
cooking 110
copper 113–14, 125–6, 165
corpuscular theory of matter 22
cosmology 153–4, 168
counterfeiting 102–3
craft skills and craftsmen 3, 35–6, 52, 56, 62, 68–9, 108, 110, 126–7
crucible 42–3, 46, 56, 67–9, 118, 125
crucible steel 115, 126
crystallino glass 121
cupel 46, 56, 68, 116–7
cupellation 40, 46, 67, 69, 115–6

Damascus mosaic 120
Das Buch der heiligen Dreifaltigkeit 165–7, 171, 173–5
Das Puch von dem Entkrist 98
Dastin, John 15, 30, 155, 159
dating of inventions 60
De aluminibus et salibus 12, 90, 96
Decknamen, codenames 37, 89, 135–6, 162–3
Democritus 7
  Pseudo-Democritus 23, 26
Demons/the Devil, cooperation with 13, 97–8, 147
descension 42, 45
descensory 67–9
Dhū l-Nūn al-Miṣrī 79
diagrams 152–60, 167–8
dialogue 134–5
dictation of text 134
dispersio technique 174–5
dissolution 43–4
distillation 15, 23, 29, 34, 40, 47, 50, 54, 65, 83, 86, 110, 125, 173
  per ascensum 40–2, 56–8, 65
  per descensum 42, 50, 56, 58, 62, 64, 67–8, 181
Dominic of Saint Proculus 101
Dominican Order 101–2, 144
Donum dei 170–1, 173–5
dramatization of chemical properties 152
drugs, extraction and use of 110
Dumas, Geneviève 147
Duns Scotus 100
dyes 123–4

Edward III, king of England 99, 147
Edward IV, king of England 175
egg, alchemical 23
Egypt 20, 23, 75, 89, 119, 162
elements of the natural world 18
Elias, Franciscan Vicar General 100
elixir theory 6, 12, 14, 26–33, 38–9, 73–6, 165
  in Latin Europe 28–31
embodied knowledge, need for 36
Emerald Tablet 83, 140–2, 145, 164
the environment, physical, effects of chemistry on 103–5
equipment used by alchemists 60–9
  practices and tools 63–9
  sources for study of 50–61
Ercker, Lazarus 112, 118
Estruç, Bernat 102
Eucharist 140
exhalations thought to be trapped underground 18–21
experimental archaeology 61
experimental science 100
Eymerich, Nicholas 13, 97, 102

Fabri de Dye, Guglielmo 147
al-Fārābī 72, 91
female authors 133
female dialogue partners 135
fermentation 110
Fez, alchemists of 95, 138
  madrasa 138

figurative imagery 152, 160–4, 167–8
early evidence from 160–2
filtration 42
Fiore, Joachim da 155
fixation 44
Flamel, Nicolas 144
Foligno, Gentile da 84
Franciscan Order 100–2
Fratelli dell'aqua vita 87
Freudenthal, Gad 88
furnaces 66–9, 114; *see also* blast furnace
Fusṭāṭ, Egypt 88

Gabricus and Beya 163
Gabriele, Mino 150
Galen, Galenic medicine 6–7, 18, 20, 29, 74
Geber 12, 38
Pseudo-Geber and the Pseudo-Geberian corpus 14, 29–30, 51, 85
Genizah manuscripts 88–91
Geoffroy, Raymond 100, 147
Géraud du Pescher 101
Gerhardus de Fracheto 97
Gershon ben Solomon 91
al-Ghawrī 139
al-Ghazālī 2, 79
Giles Du Wes 153
Giles of Rome 13
Giovanni d'Andrea 103
glassmaking 112, 118–22, 126–8, 150, 168, 173
glazes 124–5
glossaries 141
gold 5, 22, 31, 33, 125–7, 132
gold mines 97, 127
drinking of 84
Gratheus 163–5, 170
Gratian 103
Greco-Egyptian alchemy 6
Greek alchemy 4–7, 20–1, 26, 75, 134, 140, 160, 168, 173
Guido de Montanor 31
Gundissalinus, Dominicus 81, 94

al-Ḥākim, Fatimid caliph 79–80
Halleux, Robert 142
Hallum, Benjamin 162
al-Hamdānī 52, 77
al-Haṣan al-Baṣrī 79
handicraft 3
Henry VI, king of England 99, 175
Hermes Trismegistus, Hermetic 4, 7, 74, 80, 83, 93–4, 133, 140–2, 161–2, 167–8, 173
hieroglyphs 10, 53, 144, 162
Holmyard, Eric John 50
Holy Trinity, the Book of see Das Buch der heiligen Dreifaltigkeit
Hortulanus 142, 145
Hugh of St Victor 81
Ḥunayn b. Isḥāq 76–7
Ḥusayn, Ḥidāyat 50

Ibn al-'Arabī 79
Ibn al-Akfānī 73, 76
Ibn Arfa' Ra's 9, 78, 133–4, 183–4
Ibn Farīghūn 72
Ibn al-Ḥajj 80
Ibn al-Nadīm 72
Ibn Sīnā 4, 11–14, 20, 22, 27–8, 33, 42, 51, 72, 77, 82–6, 95, 140
Ibn Taymiyya 80
Ibn Umayl 9, 12, 78, 89, 132, 161–4, 167, 173
Ibn Waḥshiyya 78, 89, 135
iconography 53, 147
Ikhwān al-ṣafā' see Brethren of Purity
imagery and image-making 150–2, 159–78
Indian alchemy 4
Industrial Revolution 108
industries
case studies of 112–26
definition of 108
as precursors to later developments 127–8
instruments 67–8; *see also* equipment
al-'Irāqī, Abū al-Qāsim 9, 53, 162
Ismā'īlī lineage 79

Ja'far b. Muḥammad 138–9
Ja'far al-ṣādiq, Shiite imām 8, 79, 138
Jābir b. Ḥayyān 8–9, 11–12, 14, 22, 27–8, 38, 73–6, 89, 179
Jalāl al-Dīn Iskandar, Timurid prince 139
Jalāl al-Dīn Rūmī 79
al-Jawbarī 94
Jesuati Order 87
Jesus Christ 14, 99, 105, 140, 164–7, 170

Jewish communities 87–91
al-Jildakī 10, 75, 138, 154, 182
Johannes de Lachellis de Fontaneto 148
John I, kIng of Aragon 99
John XXII, pope 13, 86, 97, 102
John of Garland 141
John of Rupescissa 15, 31–3, 42, 86, 100–3, 147, 156
Judah Ha-Levi 91
al-Junayd 79

kabbalah 87–8; *see also* Christian kabbalists
al-Kāthī, Abū al-Ḥakīm Muḥammad ibn 'Abd al-malik al- Ṣāliḥī al-Khwārizmī 51
Khālid b. Yazīd, Prince 7–8, 99, 135, 138–9
al-Khwārizmī *see* Abū 'Abd Allāh
Kilwardby, Robert 81
al-Kindī 112
Kitāb al-ḥabīb 50
Korah, Islamic mythical figure 8
Kraus, Paul 8
Kurzmann, Peter 55

laboratories and laboratory practices 49, 59, 63–4, 69, 145
Lachellis de Fontaneto, Johannes 148
language 142
Latin alchemy 3, 12–15, 21–6, 30, 69, 80–3, 139–46, 153, 164, 170
lead 104, 115–16
  lead oxide 43, 46, 109, 119, 121–6
learning alchemy
  face-to-face learning 134
  informal learning 138
  places for 138
  with a master 138; *see also* alchemy: teaching and learning of
Lefèvre d'Étaples, Jacques 83
Lennep, Jacques van 50
Leo Africanus 95, 138
the Levant 119, 126
Libavius, Andreas 148
Liber secretorum Bubacaris 51
Liebrenz, Boris 137
Lippmann, Edmund Oskar von 73–4
literary forms 131–4
  slips in textual transmission 134

Llull, Ramon (Raymond) 15, 30, 32–3, 84–6, 94–5, 100, 144, 147–8, 155–6, 160, 168, 173, 175–6, 183
lore, alchemical 141
  bookish nature of 143
Louis II, king of Hungary 99
Lustrach, Jaime 99
lute (material) 56, 65, 68

madrasas 137–8
magic 72, 73, 82–3, 100–2
Maimonides 91
al-Ma'mūn, Abbasid caliph 134
al-Manṣūr, Abbasid caliph 99, 138–9
Mandeville, John
manuscripts 4, 52–5, 137–8, 146–9, 155, 175
Mappae clavicula 10, 66, 112, 139
Marcus Graecus 40
al-Marrākushī 132
Martin I, king of Aragon 99
Martin of Poland 103
Mary the Copt/the Jew 11, 87, 133, 182
Mary, the Virgin 166
Maryānus (Morienus) 7–8, 11, 99, 135
Maslam b. Qāsim al-Qurṭubī 9, 73, 78, 82, 95, 99, 161
mechanization 127–8
medicine and alchemy 14–15, 74–7, 81–7, 138
Medina mosaic 120
Melchior De Sibiu (Cibinensis), Nicolaus 99
Menaḥem, Magister 90
Mercury (planet) 94–5
mercury (substance) 5, 19–22, 27, 30–1, 34, 38, 75, 154–5, 162–8, 173, 178
  'mercury alone' theory 14, 23, 34, 148, 180
metal production and metalworking processes 110–18, 125–6
metaphor 105–6, 154
Meung, Jean de 143
Michael Scot 13–14, 96
mints 128
Moses, prophet 79
mosques 138
Muḥammad b. Umayl al-Tamīmī *see* Ibn Umayl

Muḥammad ibn Aḥmad al-Kātib al-Khwārizmī *see* Abū ʻAbd Allāh
Murchi, Thomas 148
Muṣḥaf al-jamāʻa 135; *see also Turba philosophorum*
mythical figures 4, 8

natural philosophy (natural science) 72, 77–8, 81–5, 140, 145, 147, 154
  alchemy seen as 77–8
New College, Oxford 121
nigromancy 103, 144
  theoretical and practical 82
Noah 8
Nürnberg ordinance 99

Obrist, Barbara 150, 154–5
occult properties 20
occult sciences 73, 82, 137
Odington, Walter 15, 155
Oldrado da Ponte 103
Oliphant, Philip 15
Olympiodorus 18–19, 23–24, 27
operative alchemy 84
oral transmission of knowledge 134–6
ore beneficiation 112
Oresme, Nicolas 96
organic materials 110–11
Ostanes 7
Ouroboros 24, 132, 171

Panopolis *see* Akhmīm
Paracelsian, Paracelsism 6, 74, 145, 148
Patai, Rafael 88
Paul of Aegina 18
Paul of Taranto 14, 51–2, 145; *see also Summa perfectionis*
Pebichius 19
Pereira, Michela 30, 101
perfumes, perfume-making 110, 131
Persian alchemy 7, 139
Peter of Zealand 83
pharmacology 74
Philip of Tripoli 142
philosophers' stone 73–5, 86, 99, 140, 165, 172
philosophy related to alchemy 78, 85; *see also* natural philosophy
physics: definition of 1
Picatrix 9, 82; *see also* Maslama
Pietro Bono of Ferrara: *see* Bonus, Petrus
pig iron 115
pigments, making of 122–3
plague 86
planets 23, 89, 132
plating of metal 125–6
Plato 4, 11, 18
Pliny 168
poetry 15, 78, 132–4, 149, 163–4, 170
pollution 103–4
Psellos 26
pseudepigraphy 4, 7, 38, 89–90, 135, 143–5
public libraries 137
Pythagoras 7, 140

Qalonymos ben Qalonymos ben Meir 91
Qānṣawh al-Ghawrī, Mamluk sultan 139
quintessence 15, 18, 31–2, 83, 157
Qur'ān 78–9
al-Qurṭubī *see* Maslama

Ramla, Israel 55
Rampling, Jennifer 147–8
al-Rāzī (Rhazes), Abū Bakr 9, 12, 25, 38–9, 43–4, 51, 62, 66, 77, 90, 122, 181
al-Rāzī, Fakhr al-Dīn 79
Raymond Lull *see* Llull
recipes 136, 145–6
rectification 41
red sulfur 79
religious institutions 99
religious scholars 78–80
Renaissance thinking 147
Reuchlin, Johannes
representation of properties and processes 167–71
Richard de Fournival 146
Richard of St Victor 81
Ripley, George 15, 31, 33–4, 160, 168
Ripley Scroll 151–2, 170–6
Robert of Anjou 91
Robert of Chester 10–11, 21, 80, 93–4
Roger Bacon *see* Bacon
Roman Empire, the fall of 126
Roquetaillade, Jean de *see* John of Rupescissa

Rosarius philosophorum 52, 148, 155
Rosinus 176; *see also* Zosimos
rosewater 40, 52, 56, 64
Ruska, Julius 38, 50
Rutbat al-Ḥakīm see Maslama

Sahl b. Bishr 73
Salimbene di Adam 100
Samanid traders 127
Samarkand 127–8
Samuel, Caracosa 90
Sanctellensis, Hugo 141
Sandal Castle 56, 64
Sasanian glassmaking 120
Savage-Smith, Emilie 55
Scholem, Gershom 87
sciences
  categorization of 71–3, 80–2
  subalternation of 81
Scot, Michael *see* Michael Scot
secrecy associated with alchemy 131, 175
Sedacer, Guillaume 101
  Sedacina 146
Senior Zadith *see* Ibn Umayl
serpent of Eden 173–4
Shams al-dīn al-Dimashqī 40, 52
Shiite Islam 79
Siggel, Alfred 136
Sigismund, Holy Roman Emperor 165
al-Sijistānī 79silver 31, 125–6, 132, 173
al-Sīmāwī 9–10
slag 113–14, 128
smelting 112–15, 128
smoke 18–19
society and alchemy 105–6
Socrates 7
soda ash 121, 126
Solomon, Gershon Ben 91
soul 19, 24–25
speculative science, *speculativi* 81, 83, 96
Splendor solis 178
Spondent quas non exhibent, papal bull 13, 97, 102
spirit 19, 24–25
Stapleton, Henry Ernest 38, 50
state institutions 99
status of alchemy and alchemists 93–6
steel 126–8
Steinschneider, Moritz 87
Stephanus of Alexandria 19, 23–4, 27
sublimation 19, 44–5, 59, 106
substances
  classification of 23–6
  composition of 18–23
Sufism 79–80
Sufyān al-Thawrī 79
sulfur 5, 20–1, 24, 75, 163; *see also* red sulfur
Summa perfectionis (magisterii) 14, 22, 38–44, 47, 51, 84, 95, 140, 143, 145, 181
Suso, Henri 87
Synesius 19, 26–7
synonymiae 141
Syriac alchemy 4, 6, 24, 26, 75

Tabula chemica see Ibn Umayl
tacit knowledge 36, 136
Tadmekka 127
talisman, word for 141–2
tanning 104–5
technical drawings 150–1
Teniers, David the Younger 63
Tepe Gawra, Mesopotamia 62
Testamentum 15, 30–2, 144, 148, 155–6
testing (and guarantees of testing) of ingredients for alchemy 37
Thābit b. Qurra 135
Theodoric de Cervia 41
Theophilus 10, 52, 112, 183
Theosebeia 135, 162
Thölde, Johann 144–5
Thomas Aquinas *see* Aquinas
Thomas de Pizan 99
al-Tilimsānī, Muḥammad b. al-Ḥājj 136
tin oxide 112, 114, 125
Tome of Images 7, 53, 162–3
tools used by alchemists 49–50, 68–9; *see also* equipment
Tower of London mint 128
trade 107, 110–12, 126–7, 146
transformation 17, 22, 27, 35–6, 110, 123, 125, 154, 170–72
  alchemy as a science of 35
transmutation of metals 2, 11, 13, 39, 73–7, 85–9, 97, 102, 140, 144, 146, 157, 173
treasure hunting 73
al-Ṭughrā'ī 9, 133
Turba philosophorum 12, 132, 135, 145, 168

Ulfbehrt swords 126
Ullmann, Manfred 8, 74–5
ʿUmara ibn Hamza 99
universities 140

Valentinus, Basilius 144–5
vapor 18–19
Venice 99
vernacular material 147, 150
Vienna, alchemists in 95
Vikings 127
Vincent of Beauvais 13, 27–33, 50–1, 81–2, 84, 100
visual metaphors 168
visualization 150
  using diagrams 152–60
Vladislav II Jagello, king of Hungary 99

al-Walīd, Umayyad caliph 120
water power 127
Wiedemann, Eilhard 50
William of Moerbeke 86
William of Ockham 100
Williams, Alan 116
women's links to alchemy 133
words
  exact meaning of 141–2
  polysemic nature of 142
workshops, alchemical 63–4, 138–9
wrought iron 115
Wycliffe, John 105

Yaʿqūb al-Manṣūr, sultan 139
Ylarius and Virgo 163
York Minster 121
Yūhīn, Indian sage 135

zinc oxide 126
zizyphus tree 73–4, 78
Zosimos of Panopolis 7, 18–19, 23–7, 50, 53, 65, 87, 134–5, 162, 176